Living with Nuclear Radiation

Living with Nuclear Radiation

Patrick M. Hurley

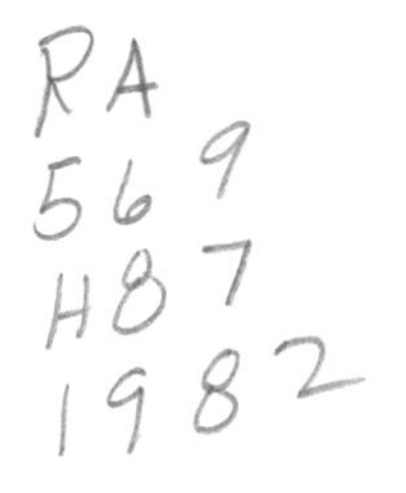

Ann Arbor
The University of Michigan Press

Published in the United States of America by
The University of Michigan Press and simultaneously
in Rexdale, Canada, by John Wiley & Sons Canada, Limited
Manufactured in the United States of America

6 5 4 3 2 1 1985 1984 1983 1982

Library of Congress Cataloging in Publication Data

Hurley, Patrick M., 1912–
Living with nuclear radiation.

Bibliography: p.
1. Radiation, Background. 2. Radiation—Physiological effect. 3. Radiation—Dosage. I. Title.
RA569.H87 1982 612′.01448 82-8351
ISBN 0-472-09339-8 AACR2
ISBN 0-472-06339-1 (pbk.)

Preface

The public concern over the construction of electric generating plants using nuclear energy is due to the possibility of a major accident and to the likelihood that nuclear proliferation in any form will eventually lead to a spread of nuclear weapons. Even the smallest of these weapons could impose threats and problems far greater than the acts of terrorism and hijacking that we have had to live with in recent years. These are valid concerns. The fact that other nations are proceeding with the development of nuclear power in an energy-hungry world only makes the decision more complex. Furthermore, a decision for or against an increase in nuclear power generation is made extraordinarily difficult by the fact that the principal alternative, the use of fossil fuels, may result in global damage to our atmosphere and its delicately balanced regional climates.

This book dwells on one factor only: nuclear radiation. It does not address the larger questions in the debate. It does not attempt to justify coal, or nuclear fuel, as the major energy source of the future. Its purpose is to clarify only that aspect of the nuclear debate that relates to dosages of nuclear radiation when everything is operating normally. One of the fears in the nuclear debate is that there will be a generally increasing level of nuclear radiation in this country as a result of nuclear power generation. Is this true? If so, to what extent, and how great will be the effect on public health? This book attempts to isolate and consider in detail only this particular aspect of the debate. A balanced and informed perspective on this subject alone should bring sharper focus on the other principal issues.

Contents

1

Introduction

This planet Earth is a radioactive waste dump. It is composed of the debris left over from the gigantic nuclear bomblike explosions of unstable stars. Fortunately the earth is old enough so that the original intense radioactivity has disappeared, but the materials of the earth still contain enough of the long-lived radioactive elements to be harmful to living organisms. This residual radioactivity occurring in the undisturbed surface materials of the earth produces a natural terrestrial background of radiation which causes a small but significant fraction of the ill health, harmful genetic effects, and abnormalities suffered by humans. Man-made increments to the natural levels of background radiation are caused by all manner of disturbances ranging from the simple tilling of soil to the mining of uranium and its use for weapons and in electric power generation. It is in the context of the harmful effects of the natural background levels of radioactivity, which we can do little about, that we should weigh and judge the added effects of man-made disturbances.

This book is dedicated to all those people who honestly debate the risks involved in nuclear power. The debate is essential to the political decision, and it is only by learning the nature and magnitude of the risks that we will be able to live with them, or reject them. Probably the most difficult task facing scientists today is that of informing the public clearly, adequately, and accurately of the facts related to technological changes. Political decisions involving a risk to the public have been with us since the first tribal councils.

Decisions to move the tribe, to fight an enemy, to show mercy to criminals, all involve risk to the society at large. However, during most of man's history the public can easily weigh the risk of these activities. In our time, a political decision, such as giving the green light to a nuclear age, involves risks that are more hidden, not readily accessible, and subject to distortion by honest people who may under- or overestimate them because of the general lack of accurate information. This book is an attempt to present an unemotional, factual, do-it-yourself account of the subject of nuclear radiations. No stand will be taken for or against nuclear power. Information that is easy to understand will be provided in some detail to help the public make educated decisions on this technical subject. Risks are weighed against gains to the society and in the light of natural risks that are with us already.

Morality today includes values to be placed on the life or health of a single individual who may be sacrificed for some common gain by a large number of individuals. Studies have been made of the actual number of dollars for which an individual would be willing to risk his life. Let us say that the specific chance of your death would be one in a million: would you take this risk for $1,000 or $10,000? Tests have shown, of course, that individuals vary widely in their decisions, depending on their ages, their state of health or wealth, or even on a sense of daredevil gambling as in russian roulette with a chance of one in six.

The results of these kinds of tests generally show that very low percentages of risk for any single individual will cause him or her to vote in favor of a political decision that will be of substantial material benefit both to the individual and to the nation at large. In a nuclear-powered society let us make the best possible estimate of the risk of death to the average individual during his lifetime as a result of the danger, including catastrophic accidents. Morally, what do we say? This risk is acceptable to me in exchange for the good to the society that the energy provides? Or, do we say I will not accept the estimated death rate in exchange for the material gain to myself. In the former case the individual places himself in the position of a winner; in the latter case he considered the plight of the loser. The difficulty of this moral question is compounded by the lack of information on counterbalancing effects such as the increased number of deaths to reduced public health from the pollution resulting from coal burning or to lowered health care and welfare in a less opulent society using extreme conservation measures.

It is by no means my intention in this book to make a decision in these matters. My object is to provide information about the natural background levels of radioactivity that have been with us since before the atomic age—levels that can be fifty times more harmful to the population than an age of nuclear power. But because we have always lived with these levels, we pay them little attention and thus express unwittingly a decision to live with them. By knowing the risk of radiation-induced illness that we have suffered throughout our preatomic history perhaps we can more easily decide on the morality of man-made additions as stated in terms of fractions of that risk. Whether or not we expand nuclear power generation, in the coming decades the major contributors to radiation exposure of the population will continue to be the natural radioactivity of the ground, the radiation coming in from space, and medical X rays, provided we do not have a nuclear war.

2

The Nature of Background Radiations

Radioactive Isotopes

Defined in a simple way, atoms are made up of positively and negatively charged particles and particles that are neutral. These are called protons, electrons, and neutrons. The protons and electrons attract each other because of their dissimilar charges, but if protons are packed together in a tight space, as in the nucleus of an atom, they will repel each other because they are all of the same charge. The nucleus of the atom contains the neutrons and protons, and is surrounded by the same number of electrons as there are protons. Therefore, the neutral atom has a balanced charge, the electrons being held in orbits around the nucleus by the attractive force of the protons. The protons would fly apart in the atomic nucleus if they were not held together with neutrons by very strong nuclear attractive forces.

Atoms are normally stable. This means that the binding forces in the nucleus are sufficient to hold the protons together. However, when the nucleus is bombarded by high-energy particles it can split into fragments of various sizes which separate from each other at high velocity. This is what is meant by atom smashing. There are other ways of breaking down, or putting together, the nuclei of atoms, yielding large amounts of energy such as we see in fission bombs, fusion bombs, or in controlled reactions that give us the energy in nuclear power plants. We are not discussing these man-made nuclear transformations in this book, except to compare their effects on human health with those resulting from certain naturally occurring transformations that happen around us at all times.

Each chemical element, such as oxygen or iron, behaves the way it does because it has a particular number of protons in the nucleus. However, there may be different species in each element in which the same number of protons is packed into the nucleus with a different number of neutrons. This does not affect the number of electrons in the atom or its chemical behavior. These different species are called isotopes. If you have too many or too few neutrons relative to the most stable configuration, the nucleus will not be stable. Isotopes that have stable configurations in their nuclei have remained in the universe, and those that have unstable configurations have undergone so-called radioactive disintegration, either quickly or slowly, depending on their degree of stability. Rapidly disintegrating isotopes have disappeared unless they continue to be produced from some source such as the breakdown of another element. Some elements are made up of both stable and unstable, or radioactive, isotopes. A radioactive isotope may be so nearly stable that it breaks down, or decays, extremely slowly. This is called a long-lived radioactive isotope. In such a case, although the proportion of the radioactive isotope is decreasing through time, there may still be some of it left from the original formation of the element before the solar system developed. In the cases of uranium and thorium there are no stable isotopes and both elements are slowly decreasing in abundance in the earth. In a large number of atoms that contain a proportion of long-lived radioactive isotopes there will be a few that are decaying and giving off radiations within any particular period of time. We can measure the breakdown of the radioactive isotopes by observing the small particles or radiations that are emitted from the nucleus during the breakdown process.

Isotopes are labeled first by the particular element and then by the number of neutrons plus protons in the nucleus.

An example is potassium-40, the isotope of potassium which has 40 neutrons plus protons in its nucleus. This isotope makes up 0.012 percent of all potassium. It is slightly unstable, emitting both beta particles and gamma rays, and breaks down at a rate by which one-half of its original atoms have disintegrated in 1.27 billion years. The time period at which half of the atoms remain is called the half-life of the isotope.

Alpha and Beta Particles

A process of nuclear disintegration in which a fragment consisting of two protons and two neutrons is ejected is called alpha decay, and the fragment is an alpha particle. The nucleus may become stable at this point, or it may still be unstable and go through a series of other decay processes until it reaches a stable state. For example, the isotope uranium-238 will undergo a series of disintegrations until it forms lead-206. Uranium-238 has a half-life of 4.5 billion years, nearly the age of the earth. Another isotope of uranium, uranium-235, has a half-life of only 713 million years (so that it is much less abundant than uranium-238) and decays to lead-207. The isotope thorium-232 has a half-life of 13.9 billion years, many times the age of the earth, and also goes through a series of disintegrations before reaching a stable isotope of lead: lead-208. These decay schemes are shown in figure 1.

In addition to the alpha decay process, a nucleus can change one of its neutrons into a proton. This occurs when there are too many neutrons relative to protons in the nucleus. This causes the ejection of a high velocity electron from the nucleus, called a beta particle, and the process is called beta decay.

Gamma and X Rays

In either the alpha or beta decay processes the nucleus may be left in an excited state with too much energy. It can lose

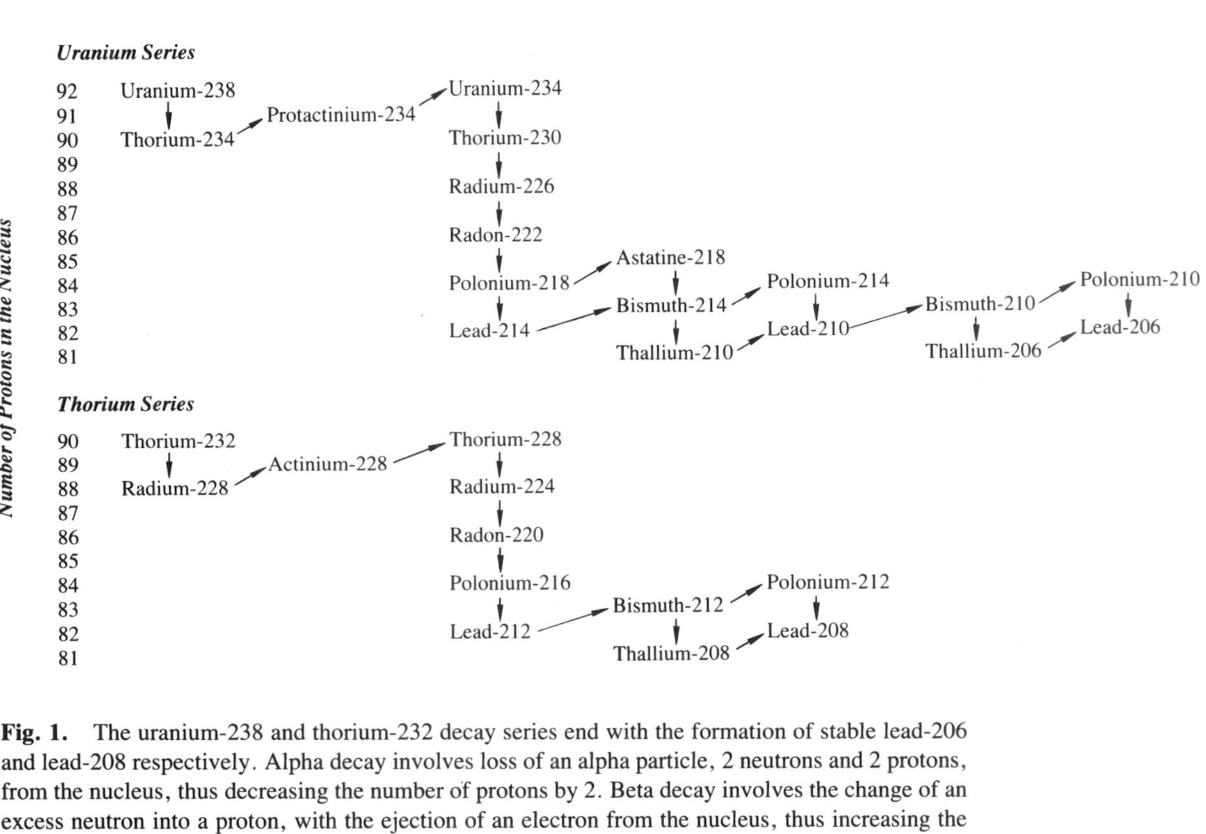

Fig. 1. The uranium-238 and thorium-232 decay series end with the formation of stable lead-206 and lead-208 respectively. Alpha decay involves loss of an alpha particle, 2 neutrons and 2 protons, from the nucleus, thus decreasing the number of protons by 2. Beta decay involves the change of an excess neutron into a proton, with the ejection of an electron from the nucleus, thus increasing the number of protons by one. The uranium-235 series is not shown.

the excess energy by emitting a form of radiation known as a gamma ray (see fig. 2). This is a pulse or packet of energy without any mass, so it is not a particle in the usual sense. It travels at the speed of light. Because the excited state of the nucleus may cover a range of different energies, the gamma rays emitted in either an alpha or beta decay process may also have different energies, up to a maximum, or may have a single energy. For example, the gamma ray coming from potassium-40 has a single energy of 1.46 million electron volts (meV), which is equivalent to the energy of an electron accelerated through a field of 1.46 million volts.

Another natural radiation is the X ray. Gamma rays and X rays are identical except that the former are generated *within* the atomic nucleus and the latter are generated by the disturbance of the electrons *around* an atomic nucleus and therefore usually have less energy than gamma rays. Most of the X rays to which a person is now exposed are man-made.

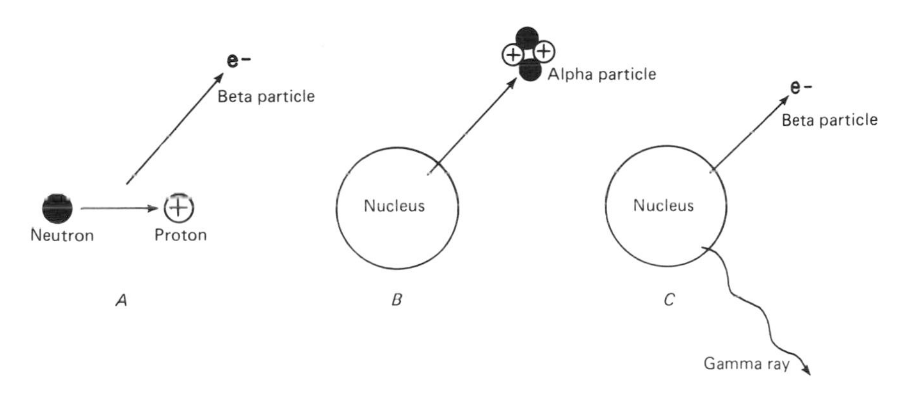

Fig. 2. Radioactive changes in the atomic nucleus. *A,* When a nucleus is unstable because the ratio of neutrons to protons is too great, one of the neutrons may change to a proton with the ejection of an electron. This is called a beta particle. *B,* In cases where the internal configuration of neutrons and protons within the nucleus is not optimal, an increased stability may be achieved by the emission of a particle consisting of two protons and two neutrons, called an alpha particle. *C,* These adjustments toward an increase in stability may require further internal rearrangements which result in the emission of photons of electromagnetic energy, called gamma rays.

3

Penetration of the Body

Alpha and Beta Particles

As mentioned in chapter 2, an alpha particle is a fragment of an atomic nucleus that is ejected with great energy from the nucleus. It does damage for a very short distance, traveling only a few centimeters in air and less than one-thousandth of a centimeter in human tissue. It can not only rip off electrons that hold molecules together, but it can also dislodge atoms directly. It is like an atomic bullet, causing intense heat and disruption to the end of its path. Radium is probably the best known alpha emitter. Because of its short range an alpha particle from radium can affect only the dead layer of the surface of the skin, and it must originate almost in contact with the body to do so. If it is inside the body, radium can do serious damage in a region very close to it (one tenth of a millimeter).

Beta particles can also dislodge electrons along their path, a process called ionization, but they do not dislodge atoms directly. In addition, they can also excite the molecular structure sufficiently by the energy imparted so that molecules along their path can be disrupted. A typical range for a beta particle in air is 15 feet, and in human tissue is 0.2 inches. Beta particles from an external source can, therefore, do damage only for a short distance into the body.

When a radioactive element is brought into the body either by swallowing a substance or breathing in a gas, it may become fixed in the body as part of the living organism. The radioactive gas radon, for example, would not normally stay in the system, but if it disintegrates while inside the body its decay products might be held long enough so that their subsequent rapid decay would take place in various parts of the system. These so-called internal sources of radioactivity can be distributed generally throughout the body, but more frequently are concentrated in certain organs or regions such as the thyroid or bones. Potassium-40, carbon-14, tritium, and the radioactive decay products of uranium and thorium are the naturally occurring sources of internal radiation. They occur in the food, air, or water that we take into our bodies daily.

Many man-made isotopes are possible. Some man-made isotopes are iodine-131, strontium-90, krypton-85, and tritium, an isotope of hydrogen. These isotopes are included in the fallout from earlier atmospheric weapons testing programs. The effluents from nuclear reactors contain radioactive isotopes of argon, krypton, xenon, and iodine, in addition to carbon-14 and tritium.

Ionization by Gamma and X Rays

The natural radiations external to the body that are most injurious to human beings are the gamma and X rays, because they can penetrate matter more deeply. If you stand on a bare outcrop of rock which contains uranium, thorium, and potassium, the alpha and beta particles developed within the rock will be absorbed by the rock and not get out, except from the surface layer which is so thin that it contains very little of these elements. In addition, the air will absorb the remainder of the alpha and most of the beta particles before they strike your body. However, gamma rays will come from deeper, 5 to 10 centimeters within the rock, so that the total amount of the radioactive elements and, therefore, the total radiations emitted, are very much greater. Also ordinary gamma rays can pass through many meters of air without much reduction in their level of intensity. This means that a person walking on the ground will receive almost the full complement of gamma radiation being emitted from the ground around him.

Gamma rays penetrating matter can travel some distance before randomly striking an electron. At this point the electron is ejected from its atom and travels through the material like a beta particle, causing the removal of other electrons in the process known as ionization (see fig. 3). The gamma ray will then proceed with reduced energy until it strikes another electron, which again causes further ionization. In this manner a single gamma ray will cause damage in a series of small regions of ionization until it finally comes to rest by having the remainder of its energy absorbed in an ejected electron (see fig. 4).

The total energy of the gamma ray is thus given up to ionization within the material, but there is a great variation in the distance traveled depending on the random striking of electrons. As a result of this variability, an average value is given for the thickness of the absorbing material through which one-half of the gamma rays will penetrate. This so-called half-value layer is thus the thickness of material in which one-half of the gamma rays are absorbed. For example, the half-value layer for the gamma rays from potassium-40 penetrating human tissue is about 12 centimeters, but would be ten times less than this through lead.

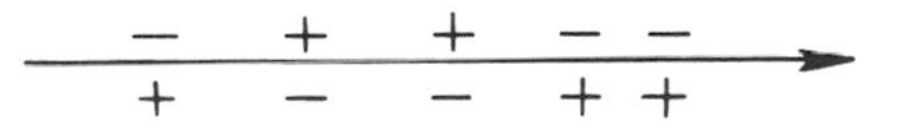

Fig. 3. Ionization. The process of ionization is the dislodging of orbital electrons by the nearby passage of a charged particle. Beta particles, which are negatively charged electrons, and positively charged alpha particles, made up of two neutrons and two protons, are the common nuclear particles causing ionization. In addition, electrons ejected from their orbits external to the nucleus by the interaction of a gamma ray, will cause ionization. Damage to living matter results from the breaking of molecular bonds, and by heavier particles, which actually dislodge atoms. Each beta particle, or energetic gamma-ray-ejected electron, will cause thousands of single ionizations before coming to a stop.

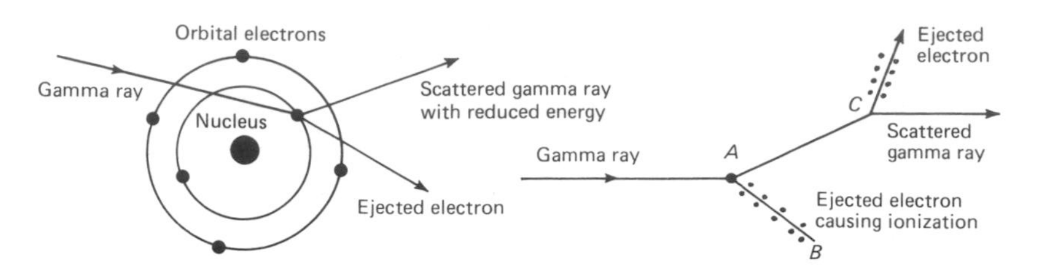

Fig. 4. Damage caused by gamma rays. A primary gamma ray, penetrating matter, interacts with an orbital electron at point *A*. It gives some of its energy to the electron, which flies off in direction *B*, causing ionization along its path. The scattered gamma ray proceeds in direction *C* with reduced energy until it interacts with another electron. This may continue until the energy of the gamma ray is reduced to the point where it is totally absorbed by the ejection of an electron from its orbit. A gamma ray entering the human body may proceed through it without interaction, may scatter out of it with one or more interactions, or may be totally absorbed within it.

Roughly speaking, the greater the density of the material, the greater is the amount of energy absorbed from gamma rays passing through a certain thickness of it. The energy imparted to a cubic centimeter of human tissue in any gamma-radiation field is considerably less than the energy imparted to bone. Consider, for example, the 1.46 million electron volt gamma rays emitted by potassium-40. For every thousand of these rays that enter the body, an average of five hundred will have been stopped by striking electrons in the first 12 centimeters of distince into the body—the half-value layer (see fig. 5). In the next 12 centimeters of penetration one half of the remaining five hundred (two hundred and fifty) will be left to proceed further, and so on.

From the half-value layer we can estimate the amount of damage done to an organ at any distance inside the body, assuming that the electrons scattered by the gamma ray do their ionizing damage over only a very short distance from the point at which they were struck by the passing ray (see fig. 6). By knowing the number of gamma rays passing into the body per minute we can estimate not only the radiation dosage imparted to the whole body, but we can also determine how much is imparted to the layers at various depths.

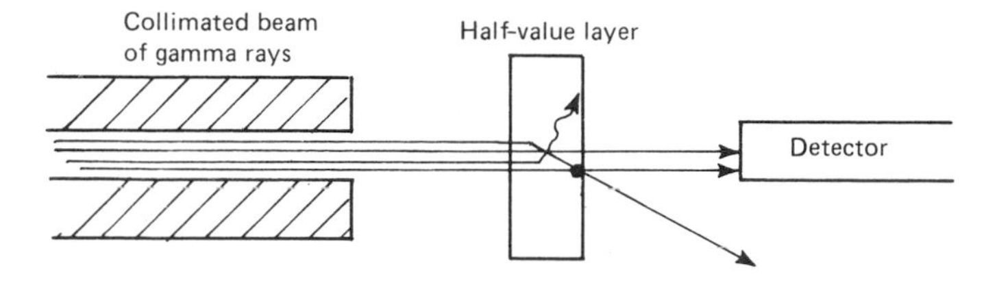

Fig. 5. Half-value layer experiment. The half-value layer is that thickness of material in which one-half of an incident beam of collimated (aligned) gamma rays is absorbed or scattered sufficiently so as not to reach a detector beyond it.

Instruments and Units: The Roentgen

Because radiations are invisible, it was not until 1895 that W. K. Roentgen discovered X rays that were produced when high voltages were applied to the electrodes in vacuum tubes. They resulted from high energy electrons leaving one electrode and striking the other of opposite polarity. The X rays were observed by the fluorescence they produced in some substances. At about the same time A. H. Becquerel, who was working with uranium ores and chemically extracting their component metals, discovered that photographic plates were darkened apparently by some form of radiation coming from these materials. This led to separations of the elements formed by the radioactive decay of uranium by M. Curie and the subsequent identification of alpha and beta particles, and gamma rays, coming from them. Soon the effect of ionization was investigated and instruments were devised which could measure it quantitatively. Eventually it was discovered that radioactive materials are all around us in small quantities giving off radiations that occur in every part of the planet we live on.

A person standing in an open space is bombarded by rays from all directions. The intensity of this radiation field may be measured by a meter or counting device which determines the number of rays entering a small volume per minute or per hour. In this type of instrument each ray is measured as a single event, not by the number of electrons dislodged or by

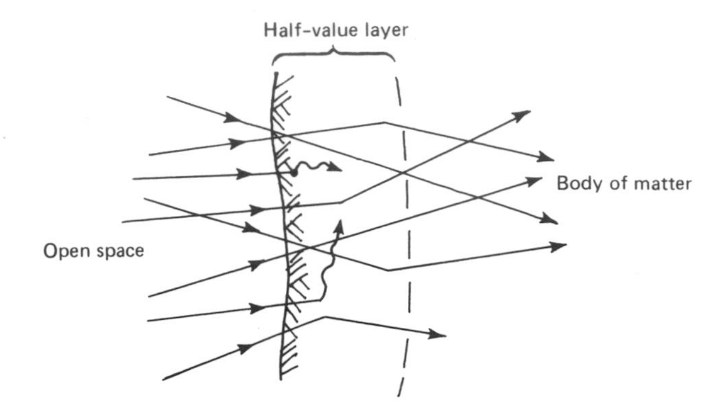

Fig. 6. When the gamma rays enter the body without orientation and scatter randomly through the half-value layer, the proportion of the gammas that get through this layer is no longer one-half. For those gammas that enter almost perpendicularly, the scattering will generally still be through the half-value layer so they will add to the number of primaries on which the half-value layer experiment in figure 5 is based. Therefore, the number of gammas getting through the half-value layer will be one-half, plus an amount specified by a "build-up factor."

the energy imparted. Commonly used instruments of this kind are Geiger-Mueller counters, or scintillation counters. Another device, known as an ionization chamber, may be used to determine the amount of ionization caused by the rays penetrating it. When each ray is counted only as a single event it is necessary to have some idea of an average energy of the rays in the radiation field to determine the amount of energy being imparted to a human body. By having some idea of the isotopes present in the ground around us, or in the air, we can estimate the average energy per ray.

When man-made X rays first came into use, their intensity was specified in terms of biological response. Later the measurement of an X-ray beam was specified in terms of the amount of ionization produced by it when passing through 1 cubic centimeter of dry air. This X-ray unit was called the roentgen, designated by the letter *R*. Because this unit became related to the damage done in human tissue it became useful to have it included in the scale of the meter in most radiation survey devices. Most of the low-level meters today have two scales: one in counts per minute (of individual rays striking the probe), and the other in milliroentgens per hour, or one-thousandth of 1 roentgen per hour (see fig. 7).

Absorbed Energy: The Rad

Different radiations cause different degrees of ionization and damage relative to their energy. Therefore, it is necessary to have another unit, called the rad, which is equal to the amount of energy absorbed by 1 gram of the material that is irradiated. One rad is defined as being the dose equivalent to 100 ergs of energy, or 62.5×10^6 million electron volts, deposited in 1 gram of material. This is particularly easily visualized in the case of beta particles, where 1 gram of material (about 1 cubic centimeter of human tissue) will absorb the total energy of the incoming radiation entering through each square centimeter. We simply multiply the energy of the beta

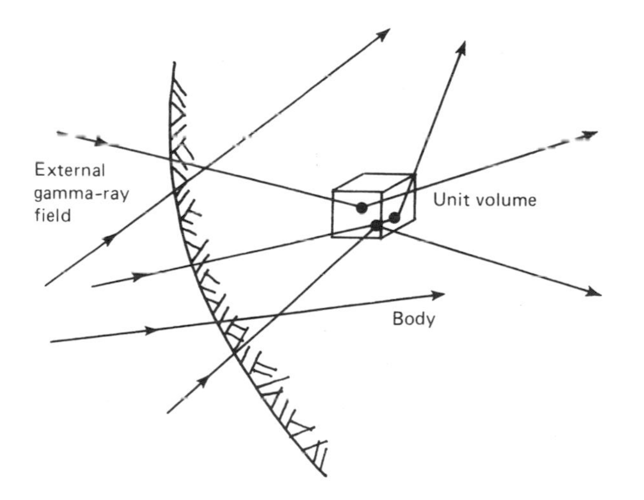

Fig. 7. The gamma radiation dose received by the human body can be visualized by replacing all unit volumes in the body by small detection probes. Each probe will record the same reading as if held in open space outside of the body, except for a reduction due to the shielding by the surrounding body. This reduction may be up to a factor of one-half.

particles times the number of them entering the 1 gram volume to get the total energy absorbed in rads, taking into account the factor used to convert the energy of the radiation in electron volts to the energy of 1 rad. In the case of gamma rays, which impart only a fraction of their energy to a single gram of material as they pass through it, the energy absorbed is the sum of the energy of the rays passing through it times the proportion of the energy absorbed.

Biological Damage: The Rem

This is still not sufficient for establishing radiation dosages in the matter of human health. It turns out that some rays, particularly the heavily ionizing particles such as alpha particles, cause up to ten times the amount of damage for the same energy imparted as do gamma or X rays. Therefore, another factor must be applied depending on the kind of radiation. When this new factor is also taken into account the rad becomes the rem: that is, a rad that is modified to allow for the kind of radiation. As far as we are concerned here most of the radiation of the natural background is made up of gamma and X rays for which there is no difference between the rad and the rem. In addition, the roentgen is so close to being the same as the rad that we can interchange roentgens and rads, and most rems, at the same value without further consideration. Therefore, in considering the health effects in natural background radiations we can convert the radiation readings in milliroentgens per hour directly to the dose rate in millirads or millirems per hour, where 1 millirad is one-thousandth of a rad. In the special case of ingested radium and other alpha-particle emitters that get into the body, the amount of damage per particle greatly exceeds the damage from an equivalent energy imparted by gamma rays or beta particles so that the dose in millirems in this case is greater than the dose in millirads.

The actual amount of energy in the ionization caused by

1 roentgen of radiation is equal to 2.58×10^{-4} coulomb per kilogram of air. This is equivalent to 5.4×10^{7} million electron volts per gram of air. This is close to the value of the rad. In fact, 1 roentgen is equal to 0.87 rad when the ionization takes place in air, or equal to 0.98 rad when the ionization occurs in body tissue.

4

Sources of Radiation External to the Body

Conversion Factors for Radioactivity in the Ground

We can divide the radioactive elements in the ground into three groups: (1) potassium, with its single radioactive isotope; (2) the elements that form the decay series from uranium, and (3) the radioactive isotopes in the decay series from thorium. For simplicity we can refer to these as the uranium series, the thorium series, and potassium. Typical concentrations of these elements in rocks of various kinds and details of the three groups are given in chapter 5.

The absorption of gamma-ray energy coming from the radioactive elements within the surface layers of the ground causes a degrading of the energies of the rays to less than their original values, on the average. In order to calculate the dosage of radiation received by a person standing on a flat surface of rock containing a certain concentration of radioactive element, one must multiply the concentration by the factors shown in table 1. As an example, a person standing on a rock containing 10 grams of uranium per million grams of rock (10 parts per million) would receive a radiation dose from the uranium and its decay series at a rate of $6.4 \times 10 = 64$ millirads per year or $64 \div (24 \times 365) = 0.007$ millirads per hour. Similarly, for a 1 percent potassium concentration in the ground the radiation above the ground from the potassium would give a dose of 13.3 millirads per year. Table 2

TABLE 1. Conversion Factors for Finding the Number of Millirads per Year Above Flat Ground Containing Each of the Principal Radioactive Elements

Content of Radioactive Element in the Ground	Radiation Exposure Rate (in millirads per year)
For each 1% of potassium	12–14[a]
For each part per million, or 0.0001%, of uranium[b]	5.8–6.8
For each part per million of thorium[b]	2.8–3.3

a. Depending mostly on the iron and magnesium content of the rock, or water content of the soil

b. This assumes that the uranium and thorium series are in equilibrium.

shows the typical contents of potassium, uranium, and thorium in various kinds of rocks, together with the uranium and thorium series derived from these elements. From this information we can see that average igneous rocks cause a background radiation dose of about 100 millirads per year, typical sandstones 40 millirads per year, typical shales about 70 millirads per year, typical limestones about 15 millirads per year, and typical granite about 200 millirads per year.

The cosmic ray background (see chap. 5), combining both radiations and particles, contributes about 30 millirads per year at altitudes near sea level.

Geometry of Source

In considering the damage done to the human system by radiations we must consider the nature of the source, its distance away, and the total time of the exposure. Sources can be at a single point, such as a vial containing a radioactive substance. In this case the radiation intensity decreases as the square of the distance away from that point source so that a person 10 meters away would receive only one-quarter of the radiation received by a person 5 meters away. If the source is

TABLE 2. Rate of Radiation Exposure to a Person Standing on Flat Ground Containing Typical Amounts of Radioactive Elements Occurring in Different Rock Types

Rock Type	Potassium (in percentage)	Uranium[a]	Thorium[a]	Average Exposure Rate (in millirads per year)			
				Potassium	Uranium	Thorium	Total
Average igneous	2–3	2–6	6–15	34	24	33	91
Sandstone	0–2	0–3	0–8	15	8	17	40
Shale	3	1–4	8–15	36	8	28	72
Limestone	0.3	1–2	1–2	4	8	4	16
Granite	3–5	3–15	12–50	40–70	20–95	35–135	95–300

a. Parts per million by weight, assuming decay series in equilibrium

a flat outcrop of rock extending in all directions uniformly, the radiation received by a person will be the same no matter whether he is 1 meter or 10 meters above the surface, except for the amount of absorption of the gamma radiation by the air between the person and the ground. If a person is in a tunnel or mine the radiation calculation using the conversion factors should be doubled.

Shielding

The effect of the air in reducing the intensity of gamma rays from the ground is as follows:

Height above Ground (in meters)	Intensity of Gamma Rays from Ground
0	100%
5	93
10	87
30	72
100	42
300	11

The intensity decreases as the height increases due to the shielding effect of the air.

Of much greater importance than air in the shielding against gamma rays from the ground are the layers of soil, humus, water, vegetation, snow, or other substances on which we walk. Similarly in a building, the foundations, the flooring, and other materials between us and the rocks of the earth's surface all act as absorbers. However, the shielding of background radiation by the structure and materials used in housing does not always offer much protection. Only about 25 percent attenuation was found for wood frame and stucco houses. Some houses made of brick or other earthy substances may even contribute to the gamma radiation by their own content of radioactive elements. Houses constructed of granite, for example, will increase the background radiation considerably over that found in the yard outside.

Time of Exposure

We have seen how the distance and the amount of absorbing material between us and the source can vary the dose we receive. Time also has to be taken into account. Obviously the longer the time of exposure, the greater the amount of damage. This seems to be a trivial statement, but unfortunately most people become greatly concerned because of a temporary exposure of an individual to radiation, whereas they forget that a much lower amount of radiation extending over a longer period of time can be as damaging when we are considering the irreparable fraction of the damage only. In the instance of the Three Mile Island accident, some people in the vicinity were greatly concerned about an exposure lasting for hours. In many of these cases the same amount of radiation damage would be occurring in these people from natural background sources anyway, within a month, or year, if the accident had never occurred. Of course without knowing the specific exposure in any one case this statement could be untrue; but the point could be made, nevertheless, for most of the fears of a nuclear accident.

5

The Natural Background Radiations from Common Rocks and Cosmic Rays

The Uranium and Thorium Series and Potassium

There are two naturally occurring, long-lived uranium isotopes: uranium-238 (99.3 percent of natural uranium), and uranium-235 (0.7 percent of natural uranium). The uranium-235 is fissionable, i.e., readily fissioned by neutrons. The uranium-238 is called fertile material, meaning that it is not fissionable, but can be converted into a fissionable material by irradiation in a reactor. The term fission refers to the splitting of a heavy nucleus into two roughly equal parts, accompanied by the release of energy and one or more neutrons. These neutrons can serve to cause the fissioning of more of the fissionable material so that the process will continue if the concentration of the fissionable material is great enough. The controlled fissioning of uranium-235 or plutonium-239 in a reactor gives rise to the energy used in the generation of electric power.

The two naturally occurring uranium isotopes, and the radioactive thorium isotope thorium-232, break down through a series of radioactive decay products until reaching a final stable isotope of lead. Most of the natural radiation comes from the decay series of uranium-238 and thorium-232, which are shown in tables 3 and 4. These tables show the half-life of

TABLE 3. Uranium-238 and the Radioactive Isotopes in Its Decay Series

		Radiation					
		Alpha		Beta		Gamma	
Isotope	Half-life	Yield	Energy	Yield	Energy	Yield	Energy
Uranium-238	4.5 billion years	75%	4.20 MeV			23%	0.048 MeV
		23	4.15				
Thorium-234	24 days			65%	0.192 MeV	4	0.092
				35	0.100		
Protactinium-234	1.2 minutes			98	2.29	0.13	0.39
Uranium-234	250,000 years	72	4.77			5	0.093
		28	4.72				
Thorium-230	80,000 years	76	4.68			0.6	0.068
		24	4.61			0.02	0.253
Radium-226	1,622 years	94.3	4.78			4	0.186
		5.7	4.59			0.007	0.26
Radon-222	3.8 days	100	5.48			0.007	0.510
Polonium-218	3.05 minutes	99	6.0				
Lead-214	26.8 minutes			100	0.72	1.6	0.053
						4	0.242
						19	0.295
						36	0.352
Astatine-218	1.5–2.0 seconds	94	6.70				
		.6	6.65				

Bismuth-214	19.7 minutes	0.008	5.52	19	3.26		
		0.011	5.45	40	1.51		
		0.001	5.27	23	1.00		
Polonium-214	1.64×10^{-4} seconds	100	7.68	9	1.88	0.014	0.799
Thallium-210	1.3 minutes			56	1.9	80	0.296
				25	1.3	100	0.795
				19	2.3	21	1.31
Lead-210	22 years			81	0.015	4	0.046
				19	0.061		
Bismuth-210	5.0 days	5×10^{-5}	5.0	99	1.17		
Polonium-210	138 days	100	5.3			0.001	0.80
Thallium-206	4.2 minutes			100	1.51		
Lead-206	stable	0	0	0	0	0	0

TABLE 4. Thorium-232 and the Radioactive Isotopes in Its Decay Series

		Radiation					
		Alpha		Beta		Gamma	
Isotope	Half-life	Yield	Energy	Yield	Energy	Yield	Energy
Thorium-232	1.4×10^{10} years	76%	4.01 MeV			24%	0.055 MeV
		24	3.95				
Radium-228	6.7 years			100%	0.055 MeV		
Actinium-228	6.13 hours			10.0	2.18	53.0	0.058
				9.6	1.85	5.2	0.129
				6.7	1.72		
				53.0	1.11		
				7.6	0.64		
				13	0.46		
Thorium-228	1.9 years	71	5.42			1.6	0.083
		28	5.34				
Radium-224	3.65 days	94	5.58			3.7	0.241
		6	5.45				
		0.4	5.19				
Radon-220	55 seconds	100	6.28			0.07	0.50
Polonium-216	0.16 seconds	100	6.77				
Lead-212				88	0.33	47	0.238
				12	0.57	3.2	0.300

Bismuth-212	60.5 minutes	9.8	6.086	64	2.25	1.0	1.81 with beta
		25.1	6.047			1.8	1.61 with beta
						2	1.03 with beta
						13	0.83 with beta
						7	0.72 with beta
						2	0.040 with alpha
						0.5	0.288 with alpha
						0.8	0.46 with alpha
Polonium-212	3.04×10^{-7} seconds	<1	10.55				
		99	8.785				
Thallium-208	3.1 minutes			100	1.80	100	2.61
						12	0.86
						86	0.58
						23	0.51
Lead-208	stable	0	0	0	0	0	0

the decay product and the yield of alpha and beta particles and gamma rays, together with their energies.

As mentioned earlier the principal sources of natural background radiation are potassium-40, the radioactive isotopes in the breakdown products of uranium and thorium, and cosmic rays. The abundances of potassium, uranium, and thorium at the surface of the earth follow well-known laws of geochemistry or geology. Potassium does not reach levels commonly in excess of about 6 percent, generally reflecting the amount of mica in the rock. Thorium does not commonly exceed 40 parts per million (1 part per million = 0.0001 percent). Uranium, however, may be concentrated in some common rocks to an extent that they might be considered as low-grade uranium ores of the future. These kinds of rocks are specifically known to geologists, but they are not generally known to the public, for the simple reason that there has been no great interest in the subject.

General Geological Distributions

In general the lighter in color the rock is, if it is a crystalline or igneous rock like granite, the higher the content of uranium, thorium, and potassium. Dark green rocks or black rocks that are volcanic in origin are generally quite low in these elements. Of course the reddish and other weathering colorations of these rocks can be deceptive, and a fresh piece is required to observe the true color of the rock. If the rock has large flakes of mica it is generally rich in potassium, and probably fairly rich in uranium and thorium. Limestones and sandstones are generally quite low in radioactivity because the former contain chemical and biological precipitates of calcium carbonate relatively free of uranium and thorium and the latter tend to have concentrations of pure quartz grains.

Continental masses float at higher elevation than the heavier rocks of the ocean floors because they contain more of the less dense minerals such as quartz and feldspar. In the same way it is a general rule that mountain belts are composed of lighter minerals and float gravitationally higher than the remainder of the continental surface. Thus there is a rough correlation between the abundance of uranium, thorium, and potassium and the elevation of the land surface. There are exceptions, of course, such as in the case of sandstones composed dominantly of quartz. If we add the effect of increased cosmic rays with altitude, discussed later, we come to a very rough generality that the higher the elevation of the land, the greater the natural background. In the United States much of our highly elevated regions are underlain by considerable proportions of granitic rocks. The Rocky Mountains are a mixture, but, nevertheless, the concentration of uranium, thorium, and potassium in them is greater on the average than in the plains and coastal belts. If one is able to identify the rocks in the area in which one lives, the information given in table 2, showing the average values of radioactivity for various rock types, would provide at least a start in estimating one's natural radiation exposure. The conversion factors mentioned earlier were used in calculating the above-ground radiation levels. Metamorphic rocks will tend to retain the radioactivity of the rocks from which they were formed. For example, the highly fissile, finer grained, mica-bearing rocks will retain the greater radioactivity of shales with various additions of sandstone and limestone. The coarse-grained banded rocks called gneisses, if light in color, may be roughly equivalent to "average igneous," or to granite, or to average sedimentary rocks composed of a previous mixture of shale, sandstone, and limestone. Generally, darker colored gneisses, as in the case of igneous rocks, will be low in radioactivity.

Equilibrium in the Uranium and Thorium Decay Series

Actually, the gamma radiation from rocks does not come from the uranium and thorium but from their decay products. When uranium and thorium have been in the rocks of the ground for a long time their decay products reach an equilibrium abundance in the rocks. This means that as much radium, for example, is being newly formed as is breaking down, and, therefore, reaches a level of radioactivity equivalent to that of its uranium grandparent. If the uranium in 1 cubic centimeter of rock has an activity of 100 alpha particles per hour, the radium will also have this rate. This is because each atom of uranium breaking down (100 per hour) will eventually turn into an atom of radium. So at equilibrium, 100 atoms of radium will be forming and breaking down per hour. The same applies to the gamma-ray emitting isotopes in the decay series. It happens that the only isotope in the uranium series that emits gamma rays that are energetic enough to get out of the rock and add substantially to the background radiation is bismuth-214. At equilibrium its gamma-ray emission will be proportional to the alpha activity of the uranium. The thorium series contains two major gamma emitters, actinium-228 and thallium-208. The gamma rays of bismuth-214 and actinium-228 range in energy up to maximum values of 2.42 and 1.64 million electron volts respectively. Those of thallium-208 (only 34 percent being formed in the series) have a fixed energy of 2.61 million electron volts. The isotope potassium-40 emits a gamma ray in 11 percent of its breakdowns, with a single energy of 1.46 million electron volts.

Despite these complexities, it can be seen that the gamma-ray production of the uranium series, the thorium series, and of potassium are all proportional to the abundance of uranium, thorium, and potassium in the rock when it is in equilibrium. Therefore, it is only necessary to measure these elements to obtain the overall gamma-ray emission from the rock. Information on these gamma emitters is shown in table 5.

One of the decay products in the uranium and thorium series is a noble gas, radon, which may escape from the rock during its lifetime. Because bismuth-214 and thallium-208 are below the gaseous radon isotopes in the uranium and thorium series, an escape of radon from the ground would mean a reduced gamma emission. There is generally a fractional loss of radon from weathering granitic rocks and from soils, but

TABLE 5. Energies, Half-lives, and Half-Value Layers for the Radioactive Isotopes That Give the Principal Gamma Rays in Natural Terrestrial Backgrounds, from External Sources

		Uranium Series	Thorium Series	
Subject	Potassium-40	Bismuth-214	Actinium-228	Thallium-208
Maximum energy (in MeV)	1.46	2.42	1.64	2.61
Half-life	1.27×10^9 yr	19.7 min	6.13 hr	3.1 min
Maximum half-value layers (in cm)				
in tissue	13.5	17.5	14.5	18.0
in rock	5.0	6.5	5.4	6.7
in air	10,465	13,566	11,240	13,953
in concrete	5.7	7.5	6.2	7.7

other rocks are quite tight. This subject is raised again when discussing man-made disturbances of the ground.

As general information, the actual number of gamma rays coming in all directions from 1 gram of rock containing 1 percent of potassium, 1 part per million of uranium and thorium in equilibrium with their decay series, is shown in table 6.

In the earth's crust as a whole there is about 3.5 times more thorium than uranium, and about 2 percent of potassium and 2 parts per million of uranium. On the average the gamma rays from uranium plus thorium roughly equal those from potassium. One cubic centimeter of rock weighs about 2.7 grams, and if one-half of the gamma rays go upward into the air, and the other half go downward and are absorbed by the rock, it can be seen that the gamma-ray emission per square centimeter from the top 1-centimeter thickness of the ground will be $2.7 \div 2 = 1.3$ times the number of gammas per minute shown in table 6, for the content of potassium, uranium, and thorium in the rock. A look at the half-value thicknesses for rock in table 5 will indicate that gammas also come from below the top centimeter. By adding up all the gammas coming from the ground and their energies, and using the values for the fraction of the energy absorbed by tissue in a person standing on the ground, it is possible to arrive at the conversion factors shown in table 1 giving the radiation dose in millirads per year.

TABLE 6. Number of Energetic Gamma Rays per Minute Coming in All Directions from 1 Gram of Rock Containing 1 Percent of Potassium and 1 Part per Million of Uranium and Thorium in Equilibrium with Their Decay Series

Gamma-Emitting Isotope	Percentage of Gammas in Radioactivity	Gamma Energy (in MeV)	Number of Gammas (per minute)
Potassium-40	11	1.46	2.02
Bismuth-214	100	0.61–2.42	0.74
Actinium-228	100	Up to 1.64	0.24
Thallium-208	33.7	2.61	0.08

As a summary it can be stated that the external gamma rays coming from terrestrial sources, mostly the ground, will give whole-body dose rates that range from 15 to 35 millirads per year for the Atlantic and Gulf coastal plains; from 35 to 75 for the northeastern, central, and far western regions; and from 75 to 140 millirads per year for the Colorado Plateau. If shielding by building structures and overlying body tissue is taken into account the average gonadal dose would be about 26 millirads per year. This dose to the reproductive organs is of principal concern in considering genetic effects harmful to the population, as will be discussed in chapter 7.

Cosmic Rays

Another important source of naturally occurring background radiation comes from cosmic rays. These are particles entering the earth's atmosphere at great velocity from outer space that are so energetic they will penetrate many feet of matter. The particles have a range in kind and abundance comparable to the elements that make up the universe, and, therefore, are dominantly the nuclei of the very light atoms such as hydrogen. They enter the earth's atmosphere from outer space at the rate of about two billion billion times per second. They have such great energies (most exceed a billion electron volts) that they cause nuclear changes in the atoms of the atmosphere. The result is that a single cosmic ray particle of high energy may produce a shower of secondary particles and radiations that can run into the millions. Many of these particles and radiations arrive on the earth's surface almost simultaneously. Because most of these secondary radiations and particles are absorbed in the atmosphere, the background level of cosmic-ray induced radiation will increase with altitude.

Table 7 shows approximate levels of radiation in milliroentgens per year at the equator, at 45 degrees latitude, at 75 degrees latitude, and at the pole, as altitude increases. For example, an average dose due to cosmic rays in Wyoming would be 75 millirads per year. At low altitudes the cosmic

TABLE 7. Cosmic-Ray Radiation Levels at Various Latitudes and Altitudes, in Milliroentgens per Year

	Latitude			
Altitude (in feet)	Equator	45°	75°	Pole
0	38	40	42	44
30,000	1,400	2,700	3,650	3,650
40,000	2,090	5,660	6,520	6,960
50,000	2,780	8,260	10,440	10,440
60,000	3,480	9,570	13,500	13,500

radiation doubles at a 6,000 foot elevation. Thus it can be seen that regular jet aircraft flying at altitudes of 30,000 feet in most latitudes could expose their passengers to a dose of cosmic radiation equivalent to 1 or 2 millirads per four-hour flight. Aircraft personnel with regularly scheduled flights may, therefore, be exposed to more than the limit permitted of 170 millirads per year (see the last section of chap. 7) from cosmic radiation alone, in addition to the dose received on the ground from background radioactivity. The absorption of the gamma radiation by the fuselage of the aircraft does not reduce the intensity appreciably. Astronauts traveling regions of geomagnetically trapped radiations may be exposed to dose rates exceeding tens of thousands of millirads per hour. Taking a weighted average of the population distribution in the United States, the average person receives a dose-equivalent rate of cosmic radiation of about 31 millirads per year.

Because the primary cosmic-ray particles interact mostly with the upper part of the atmosphere there is a great spread in the absorption of the various kinds of secondary radiation by the atmosphere. Without going into detail, the most penetrating of these secondary components that reach the ground level can penetrate up to 10 or 20 feet of rock or concrete or many floors of a tall building. The more highly absorbed parts of the secondary radiation consist largely of X and gamma rays, electrons, and positrons. This kind of radiation amounts to only about 10 percent of the total cosmic-ray ionization at sea level, but at an altitude of 10,000 feet its abundance will have increased to about 75 percent. The gamma and X rays cause the same kind of ionization effects deep within a living organism as described earlier, and the electrons and positrons cause the shallower regions of ionization in the body as described for beta particles (see chap. 3). Cosmic rays are believed to be of galactic origin and to reach the earth almost uniformly in the plane of the galaxy except for a small fraction that normally comes from the sun. During periods of solar flares in the eleven-year sunspot cycle there is a significant increase in the proportion of low energy cosmic rays coming from the sun. Sometimes the proportion reaches many times the normal level.

6

Doses from Natural Radioactivity within the Body

Isotopes of Importance as Internal Sources

Background levels of radioactivity in the environment may affect humans in a variety of ways. So far we have concentrated our discussion on the doses of radiation that are received from external sources in which the radioactive substance is in the ground, on surfaces, on the skin, or in the air around us. Gamma radiation damage is dominant in the case of external sources.

We now consider the case of internal sources where some of the most severe doses of radiation may be caused by the inhalation or ingestion of radioactive substances, or by breathing or by swallowing solid or liquid food and drink that contain radioactive substances. The radioactive gases or particles may be trapped in the lungs, potentially causing lung cancer, or be taken into the bloodstream from which they are deposited specifically in certain organs or bones. In the case of internal sources the energy derived from alpha or beta particles may be greater than the damaging energy delivered to the body by gamma rays.

When radioactive elements disintegrate within the body, giving off alpha particles, beta particles, or X and gamma rays, even the lowest energies are significant, because there is no shielding whatsoever. In this case we also have to include additional naturally radioactive isotopes that are not important in external sources. It must be remembered that the damage done by an alpha particle, or its backward recoiling nucleus, is ten times greater than an equivalent energy imparted by a beta particle or gamma ray.

The ionization caused by beta particles, and the electrons energized by gamma rays, is represented by a series of dislodged electrons along the path of the ray. But it does not necessarily include dislodged atoms. The degree of ionization along the path is roughly equivalent to the total energy. The absorbed dose is the sum of the number of beta particles or gamma-scattered electrons multiplied by the energy in each. Because a large fraction of the gamma rays will emerge from the body without interaction, the proportion of their absorbed energy must be taken into account. This proportion is typically one-third to one-half.

The radioactive isotopes that occur naturally in the human body include potassium-40, carbon-14, radium-226 and daughters, radon-222 and daughters, and directly ingested lead-210 and polonium-210. Potassium is the most important. As an example of the calculations involved, there are about 130 grams of potassium in an average adult, of which only 0.014 grams consists of the radioactive isotope potassium-40. This results in the release of beta particles at the rate of 257,200 per minute. These have a maximum energy of 1.33 million electron volts. In addition, a separate 31,790 gamma rays are emitted per minute with an energy of 1.46 million electron volts. All of the beta particles and about half of the gamma rays are absorbed in the body, resulting in doses of 18 millirems and 2 millirems per year for the betas and gammas respectively. This is in contrast to external radiations from the ground and surroundings, in which the gamma rays are the only radiations of significance.

Carbon-14 is continuously produced in the atmosphere as a result of cosmic-ray interaction with the nitrogen. This is taken into the body and expelled at a fairly constant rate so that the average amount residing in the body does not change very much. It causes about 200,000 beta disintegrations per minute with an absorbed dose equal to about 1 millirem per year.

Radium

The amount of radium-226 ingested into the body may vary up to several hundred times the average amount. Radium is released into ground waters in subsurface regions that are very low in oxygen. It is difficult to say where concentrations of radium in ground water may occur in any given region because it is rapidly precipitated, generally as a sulphate, when the waters become oxidized near the surface. Generally the only way to tell is by analysis. When radium enters the body it tends to go into the bone structure where its alpha particle radiation is damaging. There were times in the past when so-called radium springs were considered to be health spas (see fig. 8). The concept of these being good for the health was of course absurd. Fortunately many of these waters were no richer in radium than normal, despite the promotions.

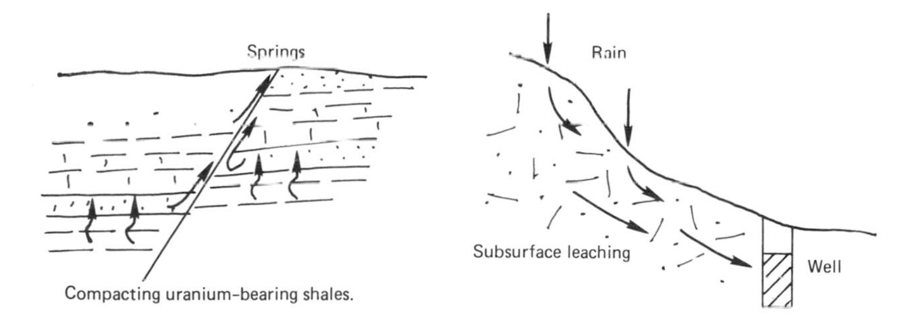

Fig. 8. Nature's waste disposal systems are generally leaky, or open. Radioactive decay products carried by fluids from compacting sedimentary rocks accumulate in the more open aquifers or faults and reach the surface as "radium springs" in extreme cases. Radioactive isotopes of uranium, thorium, radium, radon, polonium, and lead developing in the decay series of uranium and thorium are leached into all surface waters and have always occurred in our natural drinking waters.

Radium is used in industry to a very minor extent. Most of it is well controlled in storage, in medical use, or research. The majority of the industrial use was formerly in illuminated dials in watches and instruments in which the radium excites a fluorescent material to glow in the dark. Watches and clocks are still sold in this country with radium in their dials, although the use of tritium has taken over. Tritium has such a weak beta particle that it does not penetrate the glass facing of the watch or instrument. In 1971, it was estimated by the United States Public Health Service that about three million radium-bearing timepieces were sold in the United States annually with an additional large number of other devices that required permanent illumination. It has been shown that a per capita gonadal dose of a few millirads a year result from these devices, and that an individual with a luminous-dial wristwatch activated by radium could receive an annual dose of almost 5,000 millirads per year to the skin, 100 millirads to the lens of the eye, 30 millirads to blood-forming tissue, and 10 millirads to the gonads. Fortunately the use of radium is rapidly disappearing.

Radon

The decay series of uranium and thorium both include radioactive isotopes of radium. These decay to radioactive gases (radon isotopes), which in turn decay to other nongaseous radioactive isotopes before finally arriving at stable lead.

Because of this fact there is a possibility that this gaseous radon can escape into the atmosphere carrying its potential decay products with it. When inhaled by breathing, this radon and its decay products in the air will develop internal sources of radiation that are damaging to the lung and the bone. If the decay products are deposited on the surface of the land they can be taken up and ingested through the food supply, most frequently through the consumption of milk. Thus we have two possible ways of developing internal sources of natural, terrestrial radioactive isotopes: by inhalation and by ingestion. The natural occurrence of radon is mostly due to the normal escape of radon from undisturbed soil.

The intensity of radon radioactivity in any medium is usually specified by use of the curie unit. One curie (Ci) equals 37 billion disintegrations per second, approximately that occurring in 1 gram of radium.

Included in a report from the Oak Ridge National Laboratory prepared for the United States Nuclear Regulatory Commission in 1979 (NUREG/CR-0573) is a survey of all data related to measurement of the flow of radon-222 from soils in different countries. A value of 50×10^{-14} curies per meter^2 per second was selected as an average value for the natural soils in the United States. Using a surface area of 7.8×10^{12} meters^2 for the United States it is found that the total flux into the air for this country is 1.2×10^8 curies per year from natural undisturbed soils. The effect of man's technological disturbances increases this figure greatly, as will be discussed in chapter 8.

In order to estimate the health effect of any flow of radon into the atmosphere we need to use two conversion factors. The first of these gives the concentration of radon in the atmosphere resulting from a flow of radon into the atmosphere. This is determined by simply multiplying the flow rate by the factor 10^{-18} years per meter^3. For example, to obtain the average air concentration in the United States we multiply 1.2×10^8 curies per year by 10^{-18} years per meter^3 to get 1.2×10^{-10} curies per meter^3 in the atmosphere. In order to work with more usable units we can adopt the picocurie (pCi, equal to 10^{-12} curies) and state that the average concentration of natural radon in the atmosphere is 120 picocuries per meter^3.

The next conversion factor, known as the dose commitment factor, states that 1 millirem of exposure damage to the bronchial epithelium of the lung will result from a one year exposure to air containing 1 picocurie per cubic meter. Therefore, the breathing of ordinary air will cause 120 millirems of exposure per year to the lungs (see fig. 9). The effect of this on the health of the United States population is discussed in chapter 7.

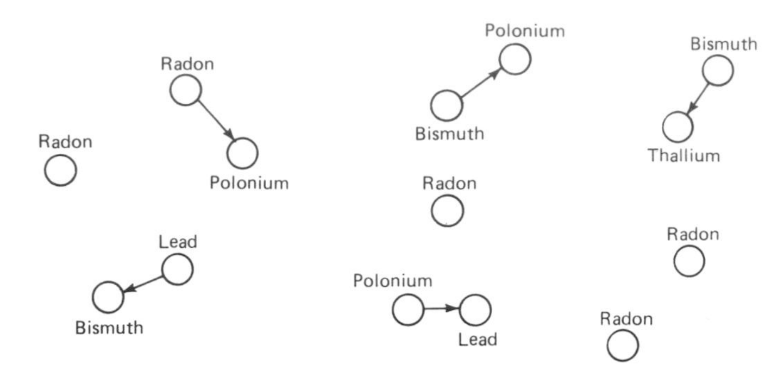

Fig. 9. Atmospheric radon and its decay products. Gaseous radon atoms forming in the decay series of uranium and thorium may escape into the atmosphere where they decay into radioactive atoms of polonium, lead, bismuth, and thallium. Although nongaseous, these atoms tend to adhere to fine particulate matter and remain suspended in the air. Inhalation of the air brings these damaging decay products into the lungs. The radon can come from the ground, the walls of a room, water from a piped supply system, or in large quantities locally from the mining and milling of uranium ores.

In addition to the natural flux of radon from the soil there is a smaller release of radon that accompanies water vapor that is evaporating from the soil and from vegetation. This so-called evapotranspiration release is estimated to cause an additional 9 picocuries per meter^3 to the atmospheric contamination.

Some rocks like granite are able to release more than normal amounts of their radon into the atmosphere. In granites up to 50 percent of the uranium and thorium may be coated on the outside of the crystal grains. When the granite disintegrates near the surface by weathering, these intergranular surfaces become exposed, and the recoil of the radon and thoron atoms releases them into the open spaces. Measurements have indicated that some weathered granites may lose up to ten times more of their radon than other kinds of rocks.

Radionuclides other than those in the uranium and thorium decay series that are of particular interest to public health are shown in table 8.

In summary, in table 9, we may see that the total dose rate

TABLE 8. Radionuclides Other Than the Uranium and Thorium Series of Particular Interest to Public Health

Nuclide	Physical Half-life	Biological Half-life[a]	Energy Radiation (in MeV)	Natural Specific Activity
Tritium	12.36 yr	9.5 d total body	0.018 beta	5–10 pCi/l in water
Carbon-14	5,730 yr	10 d total body, 40 d fat	0.155 beta	7.5 pCi/g of carbon
Potassium-40	1.26×10^9 yr	58 d	1.31 beta (89%) 1.46 gamma (11 %)	853 pCi/g of potassium
Krypton-85	10.73 yr		0.672 beta 0.514 gamma (43%)	15×10^{-6} pCi/ml in atmosphere
Strontium-90	28.1 yr	1.8×10^4 d bone 1.3×10^4 d total body	0.54 beta	No natural occurrence, man-made only
Iodine-131	8.05 d	138 d	0.36 gamma (80%) 0.64 gamma (9%) 0.72 gamma (3%)	No natural occurrence, man-made only

TABLE 8—*Continued*

Nuclide	Physical Half-life	Biological Half-life[a]	Energy Radiation (in MeV)	Natural Specific Activity
Cesium-137	30 yr	50–150 d	1.17 beta (7%) 0.51 beta (92%) 0.66 gamma (82%)	No natural occurrence, man-made only
Plutonium-239	2.44×10^4 yr	7.3×10^4 d	5.06 alpha (11%) 5.13 alpha (17%) 5.15 alpha (73%)	No natural occurrence, man-made only

a. Time required for the body to naturally reduce the concentration to one-half.

TABLE 9. Annual Equivalent Radiation Doses from Internally Deposited Naturally Occurring Radioactive Isotopes, in Millirems per year

Source	Soft Tissues (gonads)	Osteocytes	Haversian Canals	Surfaces	Marrow
Beta and Gamma Radiation					
Tritium	0.001	0.001	0.001	0.001	0.001
Carbon-14	0.7	0.8	0.8	0.8	0.7
Potassium-40	19	6	6	15	15
Rubidium-87	0.3	0.4	0.4	0.6	0.6
Total	20.0	7.2	7.2	16.4	16.3
Alpha Particles					
Uranium isotopes	12.4	7.7	4.8	0.9	—
Radium-226	0.2	16.4	10.2	6.6	1.2
Radium-228	0.3	19.0	11.0	8.0	1
Radon-222	0.4	0.2	0.2	0.4	0.4
Radon-220	0.01	0.1	0.1	0.2	0.2
Polonium-210	6	60	36	24	4.8
Total	18	110	65	44	8.5

Summary of Average Dose Rates from Natural Background

	Gonads	Lung	G.I. Tract	Surfaces	Marrow
Cosmic radiation	28	28	28	28	28
External terrestrial	26	26	26	26	26
Inhaled radionuclides	—	100–450	—	—	—
Radionuclides in body	27	24	24	60	24
Total	80	180–530	80	115	80

Source: Data from National Academy of Sciences–National Research Council 1980.

due to internal radioactivity in the human body is roughly comparable to that from external terrestrial radiation. However, it is clear that both of these sources can vary greatly, and there is a tendency for increased external radiation to coincide with considerably greater internal radiation due to intake of radon and its radioactive decay products (see fig. 10).

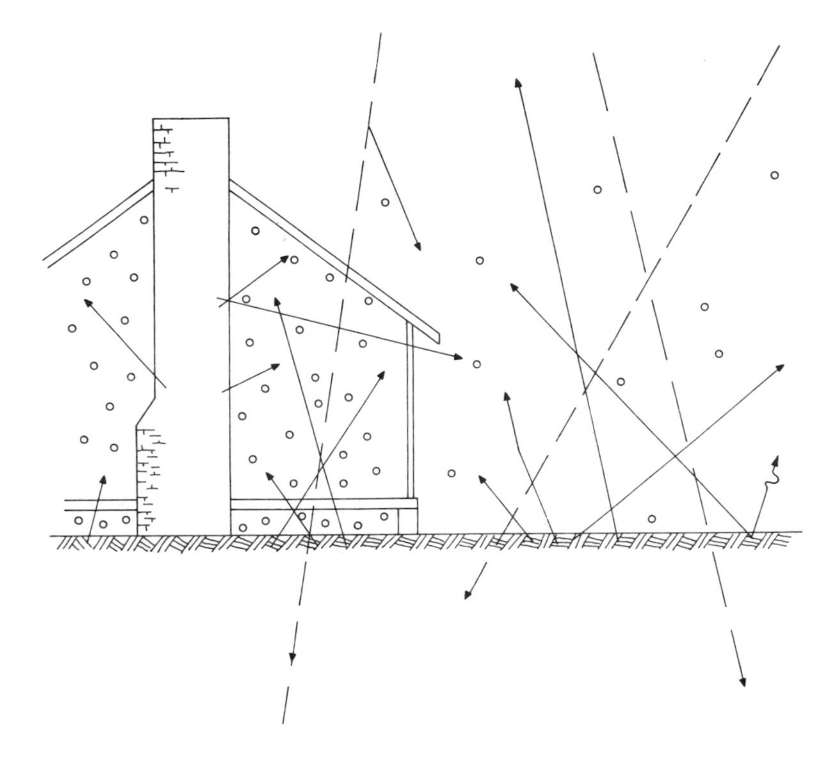

Fig. 10. Principal sources of terrestrial radiation: *dashed lines,* cosmic rays from outer space; *solid lines,* gamma rays mostly from the ground; *small circles,* atoms of radon and its decay products in the atmosphere.

7

Estimates of Risk to Human Health

Gross Overexposure

We are not concerned here with gross overexposure to radiations as might be the case in nuclear accidents or warfare. It requires from 50,000 to 100,000 millirads of this overexposure to cause effects which can be detectable clinically, such as a decrease in the number of white cells, nausea and vomiting, fatigue and bleeding. Probably one-half of the population would die after a rapid exposure to 500,000 millirads. The body can repair most of the damage in these accidental exposures if they are not too great, but there is a certain residual small proportion of damage which the body cannot repair and which may cause illnesses many years later.

The effects of radiation on exposed individuals, whether these exposures are acute, or low level and chronic, are called somatic effects, because they relate to the body of the exposed individual. Effects on the next generation, or subsequent generations, are called genetic effects. Somatic effects may be delayed for decades following an exposure, although there appears to be a tendency for the incidence of leukemia to drop off after fifteen years, and other cancers to decrease in abundance after thirty years.

Radiation exposures may cause cancers in nearly all the tissues in the human body. However, the susceptibility of different sites varies greatly, and only a few of these, such as the breast (for women), thyroid, lung, bone marrow, and some digestive organs, are sufficiently susceptible to cancer induction so that specific response data are becoming available. Leukemia is a special case of study, both because of the natural rarity of the disease and its susceptibility to induction by radiation. Naturally occurring cancers and those caused by whole-body radiation are identical, as far as is known. The incidence of both varies not only with site, but with age, sex, and possibly with the hormonal or immunological status of the individual, or whether he or she has been exposed to carcinogenic agents. The risk of latent cancer due to radiation is higher if the exposure takes place in utero, in childhood, or in the second decade of life as in the case of breast cancer in women.

Radiation-induced solid tumors are known to be more common than leukemia. Because leukemia and lung cancer are more likely to be fatal, the risk of death, as opposed to simply the incidence of cancer, is generally used in risk estimates. In other words, we generally see dose-response data expressed as the number of cases of cancer deaths, or cancer mortality, per unit of radiation dose to which a specified number of persons has been exposed. Dose-response relationships are not the same for all types of cancer, or for the type or level of radiation. Therefore, each organ or tissue must have its own risk estimate stated specifically in order to be of value in overall estimates of cancer mortality. For adults, the most important factor influencing the risk of radiation-induced cancer is age. Persons irradiated at older ages may have up to twenty times greater risk than at younger ages.

As discussed in chapter 3, the ionization due to secondary electrons traversing a tissue due to gamma and X radiation is less likely to cause structural damage to DNA than a high energy nuclear fragment such as an alpha particle. Therefore the damage per rad is less (by a factor of about ten) for gamma and X rays, than for alpha particles. The former is called low-LET radiation, and the latter high-LET. LET stands for linear energy transfer, or the average amount of energy lost per unit of particle track length. Most radiation from external sources is low-LET radiation, whereas the alpha-particle bombardment of the lung tissues due to inhaled radon and its products, for example, is high-LET radiation. These factors are taken into account in risk estimates. In the present discussion we will restrict ourselves to population averages for all cancer mortality in which these complexities do not show up.

The data on which the dose-response estimates are made includes follow-up studies on the populations of Hiroshima and Nagasaki exposed to the effects of atomic bombs in World War II; occupationally exposed populations in the nuclear industry, in weapons-testing programs, in uranium mining, and in radium dial painting; and patients who have undergone radiation treatment for diseases or who have been exposed to unusual amounts of radiation in medical diagnosis. Data on genetic effects have been derived mostly from experimentation with animals.

Estimating the Cancer Risk from Low-Level Exposures

We are concerned here most of all with chronic low-level exposures due to natural background radiation or man-made additions to these radiations. These levels are much too low to cause any immediate symptoms. For the most part the body will repair the damage that is constantly being inflicted. Again, however, there is a small proportion of the damage that is not repaired and that is believed to be cumulative. It is believed the process of ionization gives rise to highly reactive free radicals in the water of body cells that are capable of breaking the bonds in cellular DNA. Eventually the body cells descending from those cells in which the genetic code has been transformed will be present in small quantities throughout the entire system, thus acting as a latent source for cancer. The longer a person lives after radiation exposure the greater the proportion of these descendant cells that will have been formed in the system, thus increasing the likelihood of cancer to develop. In addition, and quite separate from this, there may occur superimposed effects at any time during a person's life which interfere with the normal regulation of the growth of body cells. These effects may include viral infections, exposure to chemical carcinogens, or the process of aging.

Thus we see that there can be long periods of time between the breaking of DNA molecules by ionization and the point where the body's immune mechanisms no longer can prevent the transformed descendant cells from starting an uncontrolled growth. This long time period means that estimates of cancer fatality risk cannot be completed until the studies include the entire life span following exposure.

The part of the cancer risk estimate still being studied most actively is the shape of the so-called dose-response curve at low levels of exposure. The dose-response curve describes how the statistical data on illnesses and genetic effects determined for high levels of irradiation can be extended downward to the cases where the exposure dose becomes less and less. In other words, is there some threshold of exposure below which no effects occur, or can we determine that the probability of developing a cancer is in some way related to the radiation dose even at low levels of exposure?

The 1980 report of the National Academy's Committee on the Biological Effects of Ionizing Radiation (the BEIR Report) has produced a majority report, and minority reports which disagree with the majority report. The majority report has recommended a dose-response curve providing an estimate of the cancer risks for low doses of low-LET radiation. The curve selected is the linear-quadratic model. It states that the upper limit of cancer mortality response varies linearly, or in a straight-line proportionality, with the dose decreasing to zero (the linear model). On the other hand, the lower limit of cancer mortality follows a quadratic relationship that varies as the square of the dose. Adding these two extremes together gives the committee's recommended middle-course approach: the linear-quadratic model. In this model the mortality response is dominated by the square law at intermediate levels, and by a linear decrease at low levels. For example, using the linear model, if a single irradiation of 10,000 millirads causes 1 percent of excess cancers in a sample of the population during the rest of its life, a dose of 1,000 millirads would cause one-tenth of 1 percent. In the case of high-LET radiation, such as alpha-particle bombardment of the bronchial epithelium, the linear risk estimates are believed by the committee to be better. Also, the linear model may be the best for estimates involving breast cancers.

In addition to the choice of dose-response curve, there is an uncertainty in the manner in which the cancer risk should be projected into greater age groups within a population sample. Taking an average life-table population (defined as a population with a normal distribution of ages) of one million persons, a single radiation dose of, say, 10,000 millirads will not affect young and old equally, and the projection of cancers occurring in the future will depend on how the known data are applied. There are two projection methods used by the BEIR Committee: the absolute-risk model and the relative-risk model. In the first model, if a population was irradiated at a particular dose, the risk of excess cancers would begin at some time after exposure and continue throughout the total period of expected cancers, regardless of the time of follow-up. The absolute-risk is defined as the number of excess cancer cases per unit of population per unit of time and per unit of radiation dose.

In the second, or relative-risk model, the excess cancer risk is expressed as a multiple of the natural age-specific cancer risk for that population. The relative-risk model takes account of the differing susceptibility to cancer related to age at observation for risk. If the relative-risk model applies, then the age of exposed groups, both at the time of exposure and as they move through life, is important.

Some of the risk estimates given in the 1980 BEIR Report are shown in table 10. The number of radiation-induced fatal cancers of all types due to a single radiation dose of 10,000 millirads to a population of one million persons was estimated to be between 770 and 2,250. This population has a

TABLE 10. Estimated Additional Deaths per Million Persons Due to Exposures to Low-LET Radiation

All Forms of Cancer	Absolute-Risk Projection	Relative-Risk Projection
Single exposure to 10,000 mrad		
normal expectation	163,800	163,800
excess cases: number	766	2,255
% of normal	0.47	1.4
Continuous exposure to 1,000 mrad/yr, lifetime		
normal expectation	167,300	167,300
excess cases: number	4,751	11,970
% of normal	2.8	7.2
Leukemia and Bone Cancer	**Male**	**Female**
Single exposure to 10,000 mrad		
normal expectation	9,860	8,018
excess cases: number	274	186
% of normal	2.8	2.3
Continuous exposure to 1,000 mrad/yr, lifetime		
excess cases: number	1,592	1,209
% of normal	15.0	13.4
Excess Cancer Incidence (Not Necessarily Fatal) Excluding Leukemia and Bone Cancer, per Million Persons per Year per Rad, Eleven to Thirty Years after Exposure	**Age-Weighted Average**	
	Male	**Female**
Thyroid	2.20	5.80
Breast	—	5.82
Lung	3.64	3.94
Esophagus	0.26	0.28
Stomach	1.53	1.68
Intestine	1.02	1.12
Liver	0.70	0.70
Pancreas	0.90	0.99
Urinary	0.81	0.88
Lymphoma	0.27	0.27
Other	1.52	1.64
All sites	12.85	23.10

Source: Data from National Academy of Sciences–National Research Council 1980.

normal distribution of ages (the life-table population), and the cancer mortality is given as the excess over and above the normal expectation of 164,000 cancer deaths from natural causes. The spread in numbers is due to the use of either the absolute-risk model (low figure) or the relative-risk model (high figure). Thus we see about a 1 percent increase in cancer deaths due to the 10,000 millirads of irradiation, or a total risk of death of 1 in 1,200 or 1 in 400. Estimates from the American Cancer Institute state that one-third of the population will get cancer from normal causes other than radiation, and one-sixth will die of it, as a comparison.

If the same population is exposed to 1,000 millirads per year for its lifetime, the excess cancer mortality is estimated to be between 4,750 and 11,970, or an increase in mortality of 2.8 to 7.2 percent. This is a higher chronic exposure than any expected background exposure, but might apply to long-term workers in uranium mines or mills or some other occupations in the nuclear industry. In both of these cases the linear-quadratic dose-response model for low-LET radiation was used. These figures are somewhat less than one-half of the numbers that would have been derived from the linear model.

The excess mortality due to leukemia and bone cancer from a single 10,000 millirad exposure of low-LET radiation to one million population is estimated to be 274, or 3.8 percent, for males and 186 or 2.3 percent for females. For a continuous exposure of 1,000 millirads per year for lifetime the figures are 1,592, or 15 percent, for males and 1,209, or 13.4 percent, for females. These estimates were made using the combination of linear-quadratic and linear models. However, there is some question as to whether the simple linear model may be better in the case of leukemia, in which case these figures would be approximately doubled. Note that the excess mortality refers to a percentage increase in leukemia deaths, not total deaths, and leukemia is a rare disease. Comparing the effects of radiation in the induction of cancers other than leukemia and bone cancer, it is estimated that these are two to three times more likely than leukemia for the same irradiations. Since the BEIR

Committee's report there have been persuasive arguments in favor of returning to the linear dose-response model in which there is a straight-line proportionality between dose and risk, even down to zero dose. These arguments include new data from studies of the cancer incidence and mortality from the bombing of Nagasaki and Hiroshima. Earlier estimates were based in most part on cancer fatalities. But the introduction of information from tumor registries and the use of cancer incidence in addition to cancer fatality has improved the quality of the data considerably. In addition there are new estimates related to the proportion of high-LET versus low-LET radiation at Hiroshima and new lower estimates of the gamma radiation levels in Nagasaki. These studies bring the data from Hiroshima and Nagasaki close together, and both appear to fit the linear hypothesis well. There are nearly 80,000 cases in the study altogether. The dose estimates are believed to be accurate to within about 30 percent, although there is still some continuing reevaluation. Also, because of the continuing increase in ages in the cases under study, the results are getting closer to the full lifetime estimate. Other studies are now quite consistent with the Hiroshima and Nagasaki studies, and most of them are remarkably consistent with the Nagasaki tumor registry data. The linear hypothesis seems to be emerging as the best model. Using the new data for reduced high-LET radiation in the Hiroshima case and reduced gamma radiation in the Nagasaki case, a distinguished member of the National Academy's committee, E. P. Radford, has published a report (*Technology Review,* November-December, 1981) in which he estimates the lifetime risk of death from low-level exposure is about 2.3 times the values given in table 10, and for cancer incidence about 5 times. Taking a middle position this would amount to a risk of 1 in 3,000, as the lifetime cancer death risk rate from a single exposure to 1,000 millirads. The Environmental Protection Agency (EPA) has used the value of 1 in 5,000 in its recent publications, which is well within the limits of error of all of these estimates. This figure of 1 in 5,000 is, therefore, used as the basic estimate in the remainder of this book.

Genetic Effects

The present radiation protection guides for the general public were based originally on the concept of the genetic doubling dose, or dose of radiation that will produce a number of mutations equal to those occurring naturally. The doubling dose for low-level chronic radiation is estimated to be between 50 and 250 thousand millirems with the exposure applied to the fraction of the population that is within the age range of future reproduction (namely, zero to thirty years old). Thus 5,000 millirems per year applied for a period of up to thirty years of age would eventually cause a doubling of the harmful genetic effects suffered by the population. The use of millirems instead of millirads in these statements makes allowance for the fact that some of the internal radiation affecting the reproductive organs is high-LET.

The current incidence of genetic effects is shown in table 11, together with the estimated effect of 1,000 millirems applied within the thirty-year preparental stage. It is seen that the current incidence of human genetic disorder is about 107,000 cases per million liveborn infants. After ten to twenty generations (i.e., at equilibrium) this additional radiation is expected to account for 60 to 1,100 cases, or less than 1

TABLE 11. Gentic Effects in One Million Liveborn Offspring Resulting from a Total Exposure of 1,000 Millirems per Generation Applied in the Zero-to-Thirty-Year Preparental Period.

Type of Genetic Disorder	Current Incidence per Million Live-Born Offspring	Effect per Million Liveborn Offspring	
		First generation	Equilibrium
Autosomal dominant and X-linked	10,000	5–65	40–200
Irregularly inherited	90,000		20–900
Recessive	1,100	Very few	Very few
Chromosomal aberrations	6,000	Fewer than 10	Increases only slightly

Source: Data from National Academy of Sciences–National Research Council 1980.

percent, of the natural rate. Because the normal background rate of exposure of 170 millirems per year amounts to 5,000 millirems in the thirty-year preparental period, the genetic effect of natural background would be five times that shown in table 11, or up to about 5 percent of the normal incidence. With more than 4,000,000 newborn additions to the United States population per year, the effect of normal background radiation is seen to be 1,200 to 22,000 added genetic defects per year relative to a normal incidence of 430,000. Because this value of 170 millirems per year is approximately the average background exposure and adds only 0.3 to 5 percent to the normal incidence of genetic damage, it has been adopted as the upper limit of exposure to the population that can be permitted by all man-made radioactivity.

The general "ill health" of the population resulting from genetic damage that is not specifically identifiable is estimated to be up to 50 percent of the total ill health. This leads to the conclusion that the general ill health of the population increases at the rate of about a fraction of a percent per 1,000 millirems of exposure in the thirty-year preparental period. When it is seen from the foregoing information on high natural backgrounds that chronic exposures of a few hundred millirems per year are not uncommon, resulting in several thousand millirems in the zero-to-thirty-year age range, and that this would result in a few percent increase in the general ill health of the population, the importance of the natural background can be visualized.

It is not expected that this 170 millirem per year upper limit will be approached by man-made additional radiation levels in a nuclear age. In fact, it is the plan of regulating and monitoring agencies that the additional population exposure from all man-made sources of radioactivity shall not exceed 1 percent of this, or actually less than 1 millirem per year.

Medical and Dental X Rays

X rays cause ionization in the body whether they come from natural or man-made sources. Today the doses of X radiation that affect the public health are almost entirely due to medical diagnostic and therapeutic applications, which account for 90 percent of all man-made radiation. In discussing this subject it is important to keep in mind the fact that the use of X rays in medicine cures or helps to cure hundreds of thousands of people annually, and the much smaller loss of life due to cancer resulting from X rays is amply offset by this. However, no tabulation of the radiation doses received by the population would be complete without mentioning this source, which on the average amounts to about 35 percent of the total of natural and man-made radiation.

There is a concern by the Food and Drug Administration (FDA) that the use of medical and dental X rays has been increasing to the extent that there is actually a fraction of overuse in which the risks outweigh the medical benefits. X rays are like gamma rays and are known to induce cancers and leukemia. Overuse not only contributes to the risk to the patient but adds to medical costs. Physicians have been relying more and more on X rays to protect themselves against possible prosecution for malpractice. According to the FDA the use of X rays has been increasing at a rate up to 5 percent per year, and as much as 30 percent of these X rays in some areas are unnecessary. The American Dental Association has warned against overuse, and the American College of Radiology and the American College of Obstetricians and Gynecologists has set guidelines limiting the exposure of pregnant women to pelvic X rays.

Inefficient use of the equipment, less than optimal shielding, and the fact that the radiologist is separated from the diagnosing physician in time and space so that more X rays are taken than necessary have all contributed to the overexposure. Many physicians and dentists will not supply X-ray films, no matter how recent, on request. The admission to some hospitals often requires routine X rays for both patients and employees. The practice of screening large numbers of people for tuberculosis and other pulmonary ailments and breast cancer is now on the decline, but there is much current debate on the question of whether this is advisable.

Assuming that a person's lifetime exposure should not

exceed 50,000 millirems, it was unfortunate that abdominal fluoroscopic X-ray examinations that formerly were a common practice could approach this level. Presently the upper and lower gastrointestinal series using barium are in a range less than 1,000 millirems. In some parts of the country the population receives at least one X-ray examination a year for medical or dental purposes. In the United States as a whole the rate is about 0.7 of an X-ray examination per person per year on the average. The cost of these X rays in the United States runs into several billion dollars a year, so that any increase in efficiency, or reduced usage, would provide a large saving in the cost of health care.

All persons should question the use of routine X-ray examinations and let the physician or dentist be in a position of defending this practice in any specific case. The wearing of lead aprons to shield the reproductive organs of both male and female patients should be mandatory. X rays used for the determination of stomach ulcers can still run as high as 5,000 millirems in exposure unless the equipment has been recently checked or updated. Chest X rays are generally in the range of 10 to 30 millirems, if film is used. Gallbladder X rays average 200 millirems of exposure. The very important new development of the computed tomographic (CT) scanner can take X rays in three dimensions of sections of the body. This may require 1,000 millirems of exposure, per examination.

The Bureau of Radiological Health of the FDA has estimated that the average absorbed-dose rate for the bone marrow of the adult United States population for medical diagnosis and radiotherapy is about 100 millirems per year, ranging from 50 to 150 in the spread of ages from young adults to the elderly. The total population dose from all X rays is estimated to result in about 3,000 deaths per year in this country.

On the other hand, in weighing the benefits versus the risks, most physicians feel that the present abundant use of X rays is justified. Most organs are relatively resistent to radiation. The most sensitive organs are the bone marrow, epithelium (the cells lining the skin and internal organs), the gonads, and the ovaries. There is considerable argument about the value of mammograms in detecting breast cancer as opposed to the risk that repeated X-ray exposure might offer. Some physicians advise mammography only after the age of fifty. Many who require chest X rays are asking for only one exposure of the chest, back to front, and not side views unless something shows up. Because X-ray exposure of the pelvis region in pregnant women is believed to be associated with a serious increase in childhood leukemia, this practice has been reduced.

The most hopeful new development in radiology is the use of electronic and rare-earth image intensification. Rare-earth phosphor screens placed in contact with X-ray film give off light when struck by X radiation and thus expose the film with much lower X-ray intensities than when the X rays themselves are needed to cause the exposure. A large part of the existing inefficiency in the use of medical X rays lies in the continued use of older, less collimated and controlled, and lower voltage equipment scattered throughout the country in small medical offices.

The Food and Drug Administration has assessed the doses received per average X-ray examination, for specific purposes, using modern techniques, as follows:

500 to 1,000 millirems
- Lower G.I. series (barium enema)
- Pelvimetry (to evaluate birth canal)
- Upper G.I. series (barium swallowed)
- Mammography (one breast)

200 to 500 millirems
- Lumbrosacral spine (lower spine including tip)
- Small bowel series
- Intravenous pyelogram (kidney, ureter, and bladder)
- Lumbar spine (lower back)
- Thoracic spine (middle back)

50 to 200 millirems
- Gall bladder
- Abdomen
- Ribs
- Pelvis

Skull
Hip
Less than 50 millirems
Cervical spine (neck)
Femur (upper leg)
Chest (radiographic)
Dental (whole mouth)

Recommended Exposure Limits

The need for radiation protection and regulations has resulted in a well-developed technical field named radiation protection, with training programs and handbooks. The extensive regulations protect the population down to a point where the risks are not permitted to be much greater than the normal natural background levels that humans have lived with since they evolved on this planet. However, the "normal levels" of background may be greatly exceeded in a small fraction of the total inhabited areas. We should not only be aware of these higher background situations, but we should also weigh the possible radiation exposures in nuclear accidents against the higher levels of natural radiation that people accept without question. The judgment of the acceptability of the added risk from nuclear accidents should be made with a full knowledge of how it compares with the risks to the population who are already living in natural radiation fields that may be much greater than the norm.

In the decades following the World War I, the practical use of X rays and radium had developed to a point where much better regulation was needed. As a result, in 1928 the International Commission on Radiological Protection (ICRP) was organized with thirteen members, four committees, and a backup group of experts from fourteen countries. This commission has continued in its activities ever since and has been responsible for much of the information and regulation that is used on a worldwide basis. In the United States, a government-sponsored organization was set up called the National

Committee on Radiation Protection and Measurements (NCRP) and was associated with the Bureau of Standards. Later this organization received a congressional charter and became independent, receiving support from many sources, including the government. With about 50 council members and 175 scientific committee members, this group has played a major role in establishing the regulations and making risk evaluations that have been used in this country as a quantitative base for the control of the nuclear radiation hazard.

In addition to these two organizations the United Nations's Scientific Committee on the Effects of Atomic Radiation has published many important reports. Reports have come also from the Atomic Energy Commission (AEC), the Public Health Service, the National Academy of Sciences, and the Federal Radiation Council (FRC). The Nuclear Regulatory Commission (NRC) has the responsibility for monitoring all radioactive materials related to the manufacture of fissionable materials that are manmade in the United States except for isotopes produced in devices of the nonreactor type. It is not responsible for any sources that are considered part of the natural background radiation.

The Federal Radiation Council was formed in 1960 to provide a government-wide radiation policy coordinating function and to guide the president of the United States on all matters concerning radiation and public health. However, in 1970, this function was taken over by the newly created Environmental Protection Agency, which now has responsibility for establishing federal standards and for monitoring the environment. These operations are centered in the Office of Radiation Programs of the EPA, which also is prepared to receive all public complaints or requests for further controls on environmental contamination.

Let us now look at thc various rccommended dose limits established by the National Council on Radiation Protection as given on table 12. For comparison with natural backgrounds involving mostly gamma and X rays the rem is equivalent to the rad. The millirem or millirad is one-thousandth of each of these units.

TABLE 12. Recommended Limits of Radiation Exposure

Occupational Limits of Exposure	
Whole body	5,000 mrem per year after the age of 18 and 5,000 mrem in any single year
Skin	15,000 mrem in any one year
Hands	75,000 mrem in any one year
Forearms	30,000 mrem in any one year
Other organs, tissues, and organ systems	15,000 mrem in any one year
Fetus	500 mrem in gestation period
Nonoccupational Limits of Exposure, Whole Body	
Single adult individual	500 mrem in any one year
Students	100 mrem in any one year
Population as a whole	
somatic	170 mrem average per year
genetic from medical X rays[a]	170 mrem average per year
genetic from all other man-made radiation[a]	170 mrem average per year
Emergency, Life-saving	
Individual	100,000 mrem
Body Organs	
Thyroid	1,500 mrem per year
Bone marrow	500 mrem per year
Bone	1,500 mrem per year
Near Nuclear Power Plants, Proposed Federal Regulations	
At boundary of power station	5 mrem per year
At 10 kilometers from plant	0.6 mrem per year
At 80 kilometers from plant	0.01 mrem per year
Monitoring Signs and Controls Requirement, Whole-Body Exposure	
Unrestricted area; no sign	Less than 2 mrem per hour or less than 100 mrem per week
Control of area required	More than 2 mrem per hour or more than 100 mrem per week
Radiation sign required	More than 5 mrem per hour or more than 100 mrem in 5 days
High radiation area sign	More than 100 mrem in 1 hour

Source: Some data from National Committee on Radiation Protection and Measurements 1971.

a. The sum of these genetic limits adds up to a maximum of 10,000 mrem in the thirty-year preparental period.

In considering this table we must remember that we are not concerned with cases of short-term exposure or gross overexposure which have an immediate effect on the body. We are concerned only with the small proportion of irreparable damage which can accumulate in the body and increase the chances of cancer or other problems many years later.

One of the most important recommendations proposed by the NCRP has been stated as $(N - 18) \times 5{,}000$ millirems being the limit to the long-term accumulation to age N years. In other words, no person should start working in a radiation industry until he or she is eighteen years old, and from then on he or she should be limited to 5,000 millirems per year as an average. From this basic recommendation, the limits for individuals not occupationally exposed, for the population as a whole, for fertile women, and for individual organs, have been set in accordance with sensitivity to radiation-induced somatic and genetic effects. For internal radiation exposure there is the additional concept of the accumulation of a given radioactive isotope by a critical organ, such as the thyroid for radioactive iodine and the skeleton for radioactive strontium.

Additional considerations relate to the intensity of the exposure so that an exposure dose of 3,000 millirems per quarter year is limited to 5,000 millirems for the entire year for occupational exposure. This limit of 5,000 millirems per year applies to the red bone marrow, head and trunk, gonads, and lenses of the eye. A limit of 15,000 millirems is set for other single organs. Considering the fact that radiation workers are monitored, are receiving compensation for the risk they are taking, with the possibility of planned insurance coverage and a limited period of occupation, the various committees and commissions have all generally agreed that individuals not occupationally exposed should be limited to one-tenth of 5,000 millirems per year, namely 500 millirems per year. Because the problem of genetic defects arises only in large numbers of individuals making up a population, the limiting average dose for the general population was set at one-third of this, namely 170 millirems per year, as stated earlier.

The limits for critical organs are based on the limit they would receive through whole body radiation from external sources. The ingestion of radioactive substances by the body, and their possible concentration in organs, should be considered separately so that the radiation dose is no greater than the externally imposed dose would have been. The recommended population dose limit of 100 millirems for an occasionally exposed young person is equivalent to the annual dose from the average background radiation at sea level. As will be seen in chapter 8, there are areas in the United States which may run up to ten times this level in one year. Over a period of ten to twenty years the total dose would be another ten times as great.

8

Man-made Disturbances to the Natural Background Levels

Near-Surface Concentrations of Uranium

The chemical behavior of potassium and thorium is such that they are less likely to form concentrations in the ground than uranium. Uranium can become oxidized by the oxygen of the air during the weathering of rocks. When it oxidizes, it can travel in solution in ground waters in a form that can be readily precipitated out of solution when the water is evaporated at the surface or when the water reaches a less-oxidizing environment at depth. The uranium in surface rocks is, therefore, constantly being leached and moved to another location, where it may be deposited in more concentrated forms all the way up to the great concentrations that form the uranium ores of our national reserves.

Except for unusual salt deposits, potassium has a spread in abundance in ordinary rocks and soils only up to a maximum of about 6 percent, about three times its crustal average. Likewise thorium will vary up to a few times its crustal average, except in rare cases of heavy mineral concentration in some sands. Therefore, we have to consider the special property of uranium that permits it to become concentrated near the surface, far over and above its original levels in the earth's crust.

As stated above, uranium in the surface rocks can become oxidized and travel in solution in ground waters until these flow into subsurface regions that are low in oxygen. Here the uranium may become reduced again and precipitated in concentrated form into bodies that are so rich they can be mined for uranium. Therefore, it is common to find elevated land regions in which oxidized brown and yellow rocks are yielding their uranium into solution into ground waters. These regions are surrounded by lower elevations at which the ground waters pass into the sediments of neighboring basins in which buried organic matter causes a reducing environment. Along the edges of these basins the uranium can be concentrated into a series of minable deposits of black reduced ore. It can be seen that the natural distribution of this reprecipitated uranium will not normally give any increase in background radiation because of considerable burial. However, when disturbed by mining operations these buried concentrations are brought to the surface. The same applies to all near-surface concentrations of black reprecipitated uranium below ore grade.

The same statements can apply to a lesser degree to a number of other operations in which we obtain our minerals, fuels, and rock substances for industrial use. The reprecipitation of uranium in ground waters in subsurface regions that are low in oxygen applies also to rocks carrying petroleum and natural gas, phosphate used in fertilizers, asphaltic substances, and some of the fuels such as lignites, some coals, and petroliferous deposits (tar sands and oil shale).

Coal and oil used as fuels in electric generating plants actually introduce more radioactivity into the atmosphere per megawatt-equivalent of power than nuclear power plants because of the tendency mentioned above for them to contain reprecipitated uranium, and its daughter products. Consider that all of the radioactive elements concentrated in coal and petroleum are taken from a safe disposal underground. In the process of extracting and burning these fuels they become deposited in a thin layer over the surface of the earth. It can thus be seen that this man-made disturbance to the natural radioactivity may not only be significant but increasing.

There are areas in the world where the radioactive minerals in rocks, sands, or soils have become concentrated by natural processes to such an extent that they become health hazards to the population. Extreme cases of this are the monazite sand deposits along the southwest coastal regions of India, and along certain beaches in Brazil, such as in the states of Espirito Santos and Rio de Janeiro. Fishermen, tourists, or even seekers of "health" effects supposedly resulting from radiation, or inhabitants in these areas may be exposed to fairly high levels of radiation over their own lifetime. These radioactive health spas, in which people cover themselves with black sands for their supposed curative powers, have attracted tens of thousands of vacationers. The resident populations to support these vacationers reaches into the thousands. However, it is estimated that only a few thousand are exposed to more than 500 millirems per year. In Brazil perhaps 50,000 people receive doses greater than a limit of 170 millirems a year as a result of this practice. Heavy mineral concentrates derived from sand deposits almost always contain unusually high concentrations of zircon, monazite, and other radioactive minerals that can be recovered for industrial use. Workers in mills in which these minerals are separated are probably receiving elevated doses of radiation, similar to workers in uranium mines.

Granitic rocks reach levels of concentration of uranium up to ore grade occasionally. However, the concentration of uranium in granite, which varies between a few parts per million and 1,000 parts per million, is much more commonly in the range below 30 parts per million. There is an intermediate range, however, that is not generally monitored by radioactivity surveys. This range could be the source of unusually high exposures to persons working in quarries, or in highway construction, or living on such rocks. Regions of alkaline granite bodies in the state of Minas Gerais in Brazil are so radioactive that local areas exposed by bulldozers, or farmed, range up to 25,000 millirads per year. This amount is five times higher than occupational exposure limits. Such regions match the ore-bearing regions of the world in which uraniferous materials reach the surface as stable yellow-colored minerals, and open mining operations with black ores brought to the surface, causing radiation fields up to 1,000 or 2,000 millirads per year. In southwest India nearly one hundred thousand natives are exposed to external radiation levels in this range owing to dredging and hand-working operations on the concentrations of thorium in the minerals of the ground.

The uranium concentration in natural phosphate rocks would follow the extracted phosphate in the chemical manufacture of phosphate fertilizer unless it is purposely removed. The high uranium concentration in some commercial fertilizers increases the background radiation in areas where these fertilizers are used and in rivers draining them.

Rocks that are only slightly enriched in uranium are sometimes disturbed and opened up to the surface by quarrying, farming, or road-building operations. In these cases the natural soils which are depleted in uranium by weathering, and the humus and water-soaked surface layers, may be removed, exposing the more radioactive rocks underneath. When we refer to natural backgrounds of radiation it is common to give average values that reflect the preponderance of cases in which the surface of the ground is covered by a less radioactive layer, such as soil, humus, and vegetation. These act as a shield against the radiation coming from the rocks beneath. Today, however, with the capabilities of our large earth-moving equipment more and more of the original land surface is being disturbed, exposing ground which does not have this normal shielding. Therefore, the usually quoted background values must be increased to allow for these man-made changes over increasingly large areas of the earth's surface. These man-made disturbances to the natural background radioactivity should be kept in the category of natural backgrounds and not confused with man-made radioactivity resulting from the nuclear transformations that occur in reactors, nuclear weapons, and research in high energy physics.

Uranium Ores

The geological distribution of uranium in rocks of the earth's surface has been the subject of intensive study because of interest in the reserves of uranium for a possible nuclear age. Normally the distribution of any element in the earth's surface follows a bell-shaped curve when the number of samples is plotted against the logarithm of the concentration of the element. This normal, or Gaussian, curve on a logarithmic base works quite well for elements that have a single mode of occurrence, such as substituting for other elements within the major minerals of rocks. If the element can be concentrated geochemically into enrichments in which the element forms its own compounds, these special concentrations may develop other separate distributions, or separate bell-shaped curves, at the end of the scale showing higher concentrations. Some elements such as chromium, for example, appear to have a bimodal distribution with a small bell-shaped curve well above that of the distribution of chromium in ordinary rocks.

There has been much discussion on the question of whether uranium falls under a single bell-shaped curve, or under several. The best evidence presented in a study by Deffeyes and MacGregor (1980) suggests that there is such a diversity of processes that enrich uranium that when all are put together one obtains a rather good fit to a single bell-shaped curve. It is their conclusion that the data of interest to us here can be described by a slope on the side of the bell-shaped curve that specifies a three-hundredfold increase in the amount of uranium recoverable for each tenfold decrease in ore grade. This applies not only to present ore grades but also down to levels of some not-so-uncommon rocks. A summary of typical rocks that contain abnormal amounts of uranium is shown in table 13.

The richest uranium concentrations are the vein-type deposits such as those found in the Great Bear Lake district in Canada, the Katanga area in Zaire, and Joachimsthal in Czechoslovakia. Next are the bodies of granitic rocks known

TABLE 13. Various Types of Uranium Concentrations and Estimated Above-Ground Radiation Levels Resulting from Them, Excluding the Radiation from Potassium and the Thorium Decay Series

Type of Rock	Concentration[a]	Above-Ground Whole-Body Radiation Exposure Rate from Uranium Only[b]
Average crust	1 to 3 ppm	6 to 20 mrad/y
Granites	3 to 20	20 to 130
Shales, phosphates	10 to 30	65 to 190
Black shales	30 to 100	190 to 650
Chattanooga shale	10 to 100	65 to 650
Uranium concentrations in certain volcanic rocks	100 to 300	650 to 2,000
Fossil placers, and sandstone deposits	300 to 1,000	2,000 to 6,500
Placer and sandstone ore concentrations	1,000 to 3,000	6,500 to 20,000
High grade ores	3,000 to 30,000	20,000 to 200,000
Typical ores mined in U.S.	1,500	10,000

a. Parts per million of uranium

b. In millirads per year, assuming uranium series in equilibrium

as pegmatites, together with the multiple small veins found in certain regions at the edges of ancient erosional surfaces that have been covered by later sediments. Following these concentrations there are those in sandstones formed by the reprecipitation of uranium out of ground waters, or in deposits that were concentrated in the early history of the earth by gravitational settling of heavy uranium-rich minerals in moving water. These groups form the base of the present uranium reserves of the world (see fig. 11).

As the price of uranium increases, lower grade materials will be brought into consideration, such as phosphate rock mined for fertilizer. This material often carries up to several hundred parts per million of uranium, below the grade of ores currently mined. However, the fact that it is obtainable as a by-product from fertilizer production may make it economically recoverable. If the uranium is not removed from the

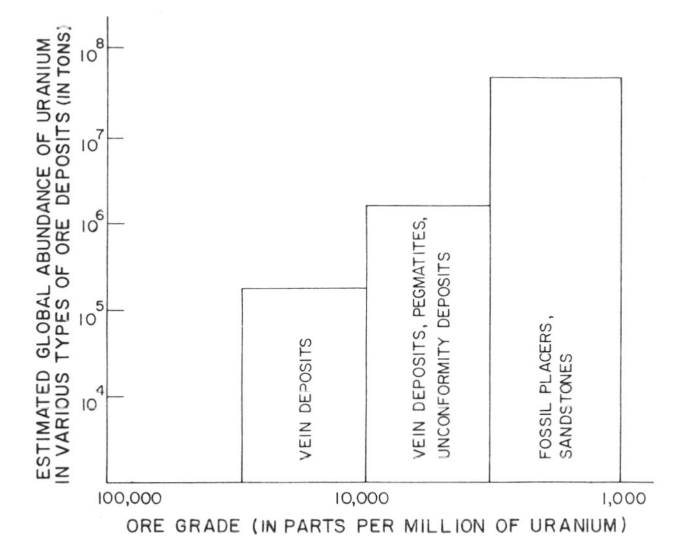

Fig. 11. Estimates of the total available uranium in the different types of uranium ore deposit. Although the content of uranium varies greatly within each deposit, the average ore grades can be specified as shown. (Data from Deffeyes and MacGregor 1980.)

fertilizer it becomes a hazard, because it is handled by farmers and spread on the surface of the earth where its radiation will be added to that of the ground. The uranium in phosphate fertilizer can be a concern; or if separated from the phosphate, an important supply of uranium. It has been shown that by-product uranium from phosphate mining alone might supply a large proportion of the fuel for the nuclear power plants in the United States if the growth of nuclear power is not expanded.

At levels below the present economic grade the geochemical processes that concentrate uranium can affect large areas of the land surface. Certain volcanic ash deposits are so permeable and release their uranium so readily that spotty concentrations of uranium can be found in broad areas in which these ash deposits occur. Black shales of marine origin, quite abundant in nature, generally contain more than average amounts of uranium. Some of these, such as the Chattanooga shale, may contain from 10 to 100 parts per million. This particular formation runs through several states in the Appalachian region and westward. It is being tested by state geological surveys as a possible ore of the future. A somewhat similar formation in Sweden is already being utilized.

Granites have been mentioned earlier, but it is not generally known that some granites are so rich in uranium that they can be mined. Notable occurrences in southern Africa and Brazil have led to a broad program of prospecting in other countries. Normally granites contain about 10 parts per million of uranium, but as in all distributions the upper third or so of all granites are higher than this.

Uranium mining today includes ore grades down to 1 thousand parts per million which are blended into higher-grade materials to make up a usable grade of ore for the mills. In the future there may be mining of ores down to as low as 500 parts per million, but without the use of breeder reactors this is doubtful. Discoveries of very large deposits in Australia, which early indications show to contain more uranium than has been produced in the United States so far, together with the bringing in of uranium from phosphate mining and the reuse of some of the uranium stockpiles in the form of obsolete nuclear weapons, could help to keep a sufficient supply coming in for the immediate future without depending on new major discoveries in the United States. The present proven reserves in the United States will supply about 100 reactors for their lifetime.

Doses of External Radiation Due to Uranium Mining

The effect of uranium mining on human health can be estimated in terms of either the external radiation from gamma rays, or the internal radiation from inhaled or ingested radioactive isotopes, mostly alpha-particle emitters. The external sources are considered first.

Assuming no recycling of uranium or plutonium, a 1,000 megawatt-equivalent power plant at 80 percent capacity consumes 27,500 kilograms of enriched uranium per year. This requires 167,000 kilograms of natural uranium, or, converting to more commonly used units, 432,000 pounds of uranium oxide (U_3O_8). Uranium oxide consumption is, therefore, 0.0617 pound per megawatt-hour (1,000 kilowatt-hours). The total nuclear power generation in 1979 was about 270,000,000 megawatt-hours. Excluding start-up inventory for new plants, the 1979 consumption of uranium oxide was about 8,300 short tons (2,000 pounds). If all the plants being built or planned come into operation, the annual consumption will be about three times this, or 25,000 short tons per year.

Considering only the worker engaged in the extraction and processing of uranium from the ground, we can estimate the risk of external radiation from the average uranium ore. Let us say, for example, that the price of $40 per pound of uranium oxide (yellowcake) carries a 25 percent labor cost and that at 25,000 tons per year the annual payroll would be about $500,000,000. This would be equivalent to about 25,000 workers, most of whom would be close to the stream of uranium-bearing materials from the mines through the mills and processing plants. The average grade of ore mined today contains about 1,500 parts per million of uranium. Using the conversion factor given earlier, a worker would receive a radiation exposure of 2,000 millirads per year if working in open pit mines or mills, or twice this if working underground. This approaches the 5,000 millirads per year of occupational exposure that is the regulation limit. For a group of 25,000 persons working for twenty years this exposure would result in an estimated total of 1,500,000 whole-body person-rems, or 300 cancer deaths, and a doubling of harmful mutations in their offspring.

In addition, the annual release of 350,000 curies of radon per year from dry tailings ponds and accumulated waste will add to local gamma radiation from the ground as its decay products are brought down to the ground surface by rain or settling dust. However this effect appears to be slight.

Internal Radiation Due to Radon and Radioactive Particulates

Almost all man-made disturbances of the soil and the natural ground waters increase the escape of radon into the atmosphere, and add, therefore, an important atmospheric contaminant which is generally little understood by the public. This contaminant has been referred to as technologically enhanced natural radiation (TENR). In order to study this TENR contamination we should keep it separate from the natural radiation background due to radon and its decay products coming from undisturbed soil. As given earlier (chap. 6) we saw that the average natural flow of radon from the soil in the United States was 50×10^{-14} curies per meter^2 per second. This resulted in a total input of radon into the atmosphere of 1.2×10^8 curies per year. Using the conversion factor of 10^{-18} years per meter^3 to obtain the average air concentration from this flow rate we find the natural background value of 120 picocuries per meter^3. As stated earlier in chapter 7 this will cause about 120 millirems of exposure per year to the lungs in the average adult. We can then use the estimated risk rate of cancer death of 1 in 25,000 resulting from 1,000 millirems of exposure of the lung in order to estimate the total number of deaths in the United States population resulting from natural radon and its decay products in the air. The population of the United States in 1978 was 218,000,000. Thus the radon from natural undisturbed soil will result in an estimated annual population dose of 2.6×10^7 person-rems, or a lung cancer death toll of 1,000 persons per year if all of the assumptions and limited data relating dose and death rate are taken to be valid.

Estimates of the man-made additions to the natural atmospheric radon, the TENR, have been made by the Oak Ridge National Laboratory (U.S., Nuclear Regulatory Commission 1979) and the Environmental Protection Agency (1979). These include separate investigations of all of the important sources of radon in the atmosphere, as listed in table 14.

During the earlier years of uranium mining in this country the health regulations for this industry unfortunately fell

TABLE 14. Estimated Nonoccupational Population Exposure and Risk of Cancer Death in the United States from Radon and Radioactive Particulates Released to the Atmosphere, 1978

Source	Estimated Annual Release (in Ci/year)	Estimated Population Dose (in person-rem to lung)	Equivalent Predicted Deaths per Year
Natural Sources			
Natural soil	120,000,000	26,000,000	1,040
Evapotranspiration	8,800,000	1,900,000	76
Technologically Enhanced Sources			
Building interiors	28,000	36,000,000	1,440
Radioactive particulates in effluents of coal-fired power plants[a]		4,750,000	190
Tillage of soil	3,000,000	680,000	27
Uranium mining industry	350,000	65,000	3
Phosphate mining and manufacture	101,000	23,000	1
Ground water treatment plants	12,000	100,000	4

Source: Data from U.S., Nuclear Regulatory Commission 1979, and Environmental Protection Agency 1979.

a. Radioactive particulates include all radioactive isotopes in the breakdown series of uranium and thorium that are airborne and inhaled as fine dust or smoke particles, and ingested through the food or water supply.

between two administrative agencies: the federal government and the state governments. The AEC did not have control over uranium mining and left the regulation of the industry to state-controlled health agencies. The result was that in the early years of uranium mining the mines were not vented sufficiently, and the unfortunate high incidence of lung cancer from radon was a disaster that could have been prevented if any reasonable care had been taken. The experience of mine workers in Europe was well known; scientists had pointed out the danger, and the industry was not unaware of radon emanation into the underground workings. However, a combination of circumstances that included unusually difficult and dangerous mining conditions in the sandstone-type uranium deposits, the slowness of the state legislative process, and the rapidity of the demand for uranium for weapons production following World War II all combined toward overlooking this hazard. Now the uranium mines are vented, and the workers are monitored for internal doses from radon and dust inhalation. This has done much to reduce this hazard.

Uranium mining, milling, and waste impoundment generate about 350,000 curies per year, equivalent to an average of about 0.3 picocuries per meter^3 in the air. However, this average is made up of high regional concentrations in areas of sparse population so that the health effect is reduced. The expected total cancer deaths per year for the entire nonoccupational population from this source is about three. This figure does not include workers in the mining industry, nor does it include the effects of external gamma rays from fallout of radon daughters or other radioactive isotopes inhaled or ingested with dust or with the food chain.

Information on coal-fire electric power generation and its radioactive effluents is given in tables 15 and 16. New plants with improved controls on emission are less hazardous than older existing plants. By far the greater problem lies with plants in urban areas. The EPA estimate of 0.7 curies per year from a 135-megawatt model coal-fired unit includes not only the radon release during the burning of the coal but also a continuing release of 20 percent of the radon from 4 percent

TABLE 15. Radioactive Isotopes in the Particulate Emissions from the Stacks of Typical Older and Newer Coal-Fired Power Stations in Picocuries per Second

Radioactive Isotope	510-Megawatt Station, Older Example	450-Megawatt station, Newer Example
Uranium-238 Series		
Uranium-238	1,180	278
Uranium-234	1,175	350
Thorium-230	74	199
Radium-226	257	202
Lead-210	1,380	734
Polonium-210	3,326	698
Uranium-235 Series		
Uranium-235	60	19
Thorium-227	—	21
Thorium Series		
Thorium-232	39	73
Thorium-228	35	85
Total	7,526	2,659

Source: Data from Environmental Protection Agency 1979.

of the flyash that escapes into the atmosphere. The average thorium content of coals is about 12 parts per million, so that its products add considerably to the total. This figure leads to an estimate of 500 curies of radon per year from 250 power stations containing the equivalent of 800 of the 135-megawatt model units.

The effect of radon release in coal-burning power plants is, therefore, trivial. However, this is not the case with the radioactive isotopes in the uranium and thorium decay series which occur in the fine particulate matter and smoke coming out of the stacks (see fig. 12). Tables 15 and 16 contain data that suggest the cancer death rate in the United States may be around 200 per year from this cause. It is interesting to note from table 14 that phosphate mining and manufacture, because of its magnitude, is almost comparable to uranium mining in its health effect.

An interesting release of atmospheric radon comes from

TABLE 16. Health Effect of Radioactive Particulate Emissions from a Coal-Fired Electric Power Station of 500-Megawatt Capacity in an Urban Area, with 80 Percent Particulate Control Efficiency Using Electrostatic Precipitators, Burning Bituminous Coal Containing 1.9 and 5 Parts per Million of Uranium and Thorium Respectively

Organ	Population Dose (in person-rem per year)	Deaths per Year from 100 Plants
Lung	19,000	76
Bone	11,000	37
Kidney	5,200	18
Liver	4,600	9
Thyroid	5,500	0.5
G.I. tract	3,700	7
Other soft tissue	5,700	28
Total		175

Source: Some of data from Environmental Protection Agency 1979.
Note: There is an equivalent of about 100 such stations operating in the United States. The remainder are newer and better controlled.

water supply treatment plants. Radon is released in the aeration process. EPA figures are specific for radon release per model treatment plant, but no estimate is presented of the total plant capacity of the United States. Assuming several thousand model plant equivalents, the risk of cancer from water treatment exceeds other industrial operations except the generation of electricity from coal.

By far the greatest health hazard comes from the increased radon concentration in the interiors of buildings (see table 14). The principal sources are the flow of radon into the building from the disturbed soil underneath it, the escape of radon from water used for domestic purposes (because the water builds up a content of radon from its underground storage reservoir), and the radioactivity of the building materials used in the construction. Measurements in the New York City area showed radon concentrations averaging 800 picocuries per meter^3, seven times higher than outside. Because most individuals spend about 70 percent of their time indoors this fact greatly increases their exposure.

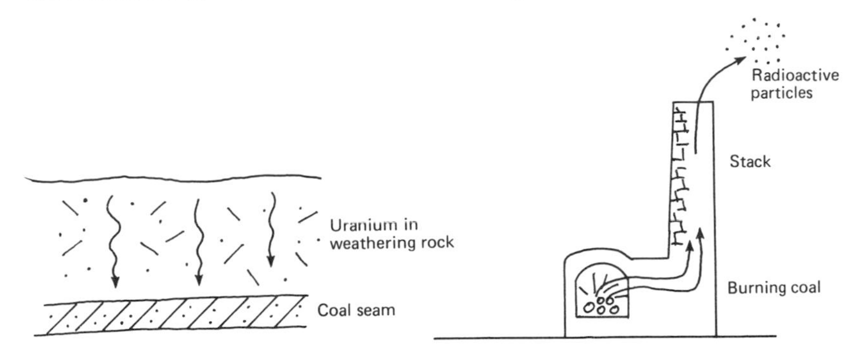

Fig. 12. Radioactive particulates from burning coal. Uranium is oxidized and leached out of overlying soils and deposited in oxygen-free coal seams. When burned in coal-fired electric generating plants the uranium and its decay products go partially up the stack, adhering to smoke and ash particles. The radon is included in the gases and its decay products are subsequently also added to the particulate matter. When inhaled the alpha-particle activity of these effluents is damaging to the lungs. The adverse effect of this on public health is greater for a coal-fired plant than for a nuclear plant of equal generating capacity.

Radon is released inside a house from the domestic water supply by faucet aeration and the use of detergents producing suds. The use of 1 cubic meter of water per day per household with an average concentration of 700 picocuries of radon per liter gives rise to an estimated 1.5×10^7 person-rems of additional exposure from this source, and 600 deaths per year. Another release comes from gas ranges and domestic heaters.

Estimates of the radon coming into typical American homes from the floors and walls can be made using a low and a high case as examples with probably the average value falling in between. The low example would be walls of cinder block or light concrete and a floor of heavy concrete, and relatively radon-free ceiling materials. In this case about 90 picocuries per meter^3 is estimated to be the additional radon contamination in the air coming from the walls and the floor. Using an estimate of 74,000,000 such dwellings give a population exposure of 5.7×10^6 person-rems and 230 deaths per year. The higher example is assumed to have a wood exterior with interior walls and ceiling covered by by-product gypsum board, which is not much used in the United States but is coming into use in other countries. This material is quite radioactive. If it comes into use extensively in the United States (say up to 40 percent of all dwellings), it would result in an exposure of 6.7×10^7 person-rems per year, or about ten times more than the lower example.

An important release of radon comes from farming or tillage of agricultural land. Estimates have shown that the working of land surfaces for agricultural purposes increases the natural radon flow out of the soil by about 13 percent. Since 3.8×10^8 acres of arable land is tilled in the United States (about 20 percent of the surface area of the country) the enhanced radon flux will be 3.1×10^6 curies per year. This will result in an increased radon concentration of the atmosphere by 3.1 picocuries per meter^3 and a population exposure of 6.8×10^5 person-rems.

On a local scale the release of radon from the ground in the Florida phosphate districts has been much more hazardous. The EPA risk assessment (*Federal Register* 44, no. 128 [July 2, 1979]) projects that over a seventy-year (normal lifetime) period, exposure to the estimated 14,000 persons residing in approximately 4,000 Florida homes existing on phosphate lands, would result in about 150 additional deaths from lung cancer. This is an additional 35 percent incidence above the normal rate. In 600 of these homes the radon content is so high that the risk of lung cancer is increased from two to four times the average risk to a member of the United States population.

In the summary of these technologically enhanced sources and resulting population exposures given in table 14, it should be remembered that these doses are from internal sources only, which make up less than half of the total radiation exposure coming from natural, terrestrial radioactive materials. The principal part of the average individual's background radiation exposure comes from external gamma rays from the ground and cosmic rays from space. Surprisingly it is seen in table 14 that by far the greatest technologically enhanced source of radon comes from building interiors. The estimate of

3.6×10^7 person-rems to the lung coming from this source would be equivalent to about 1,000 deaths per year in the United States from cancer of the lung. It is clear that wallboard materials high in radium, and thus with greater radon fluxes, should be avoided by the building industry. The cost saving in the use of by-product gypsum, for example, is not great enough to warrant the additional risk.

Adding up the death rate in the population due to cancer of the lung resulting from the doses of high-LET radiation (mostly alpha particles and recoiled nuclei) from radon and its daughter products, we obtain an estimate of about 3,000 deaths per year from this cause. Actually, recent figures are higher, to the extent that perhaps 20 percent of fatal lung cancers among nonsmokers result from inhaled radon. Smoking is credited with about 150,000 deaths per year, for comparison.

9

Man-made Radioactivity

Comparison of Risks in Electric Power Generation

Throughout the history of atomic energy there have been highly restrictive policies established regarding safety to the environment. These policies would have given us ample protection if it were not for the sudden onslaught of the weapons testing programs. During the 1950s the United States, Soviet Union, United Kingdom, France, and China carried on weapons tests that contaminated the entire surface of the earth to such an extent that concern from all nations finally caused an international agreement prohibiting further testing.

A series of studies in several countries investigating the effects of radiation appeared through the 1960s and early 1970s and paved the way in large part toward present regulations and assessment of risks. These studies are exceedingly complex; they involve the biological and physical sciences, genetics, inorganic and organic chemistry, meteorology and oceanography, and medicine. These studies also led the way for the later intensive work on environmental pollutants, such as insecticides, industrial chemicals, automobile exhaust gases, and studies of the complex ecological pathways.

In addition to nuclear power generation we should remember that the arsenal of fissionable materials locked up in our program of military readiness overshadows the use of such materials for peaceful purposes. These are not constantly being burned up producing radioactive waste products, but require the same mining of uranium, milling, processing, and enrichment as in the case of power generation. Of course in the event of nuclear war this inventory would be expended in the worst possible fashion into the environment without challenge or regulation. In such an event if populations are to survive at all, it is even more essential that the average citizen has some awareness and education in the subject of nuclear radiation.

The risks outlined in this book must not be considered in absolute terms. Life always has had and will have its risks. For example, the automobile accident death rate of 1 in 4,200 per year for the United States population (52,000 individuals per year) is an acceptable risk because of the benefit of driving back and forth to work, to shop, and to have vacations. The risk of an extended shortage of energy is very real. It would reduce the quality of life and individual freedom. A decrease in health services, public safety, food production for the world, transportation, communication, and housing comfort would also mean life shortening, illness, crime, unemployment, and even invite aggressions from other nations, leading to war.

There has been much discussion on the relative merits of coal versus nuclear fuels for the generation of electricity. Unfortunately the present information on the dangers to public health in each case can be used to distort the comparison in one direction or another because of great differences between past performance and future improved methodology, and in estimates of the health effects at low levels of toxic emissions. There is an enormous difference between the health impact of older coal-fired electric generating stations in densely populated urban areas and newly planned additional capacity with costly emissions controls, sited in rural or more remote locations. By using data on the performance of existing older plants it would appear that coal is very much worse than nuclear power generation in its impact on public health, especially if a linear decrease in health hazard is used as a basis for estimating the effect of diluted effluents containing sulphur oxides. A comparison considering only new coal-fired capacity is much more favorable.

Some amount of detail is necessary in order to present an accurate picture. There is a great difference between older plants and new plants in their removal of flyash, or particulate matter from their stack emissions by standard devices such as electrostatic precipitators, and by much more expensive control systems involving lime scrubbers. There is also a difference in the health hazard resulting from using different kinds of coal. Coal with 3 percent sulfur, for example, causes several times the numbers of deaths from sulfur oxides and suspended sulfates than low-sulfur coals. The aggregation of the coal fed into the burners affects the relative distribution of radioactive elements between the flyash and the slag. The siting of the coal plant in the center of a densely populated urban area as opposed to a more rural location with a lower population density in the predominant downwind direction also makes a great difference. We, therefore, come up with the numbers shown in table 17 in which the health hazard for a unit plant varies by a factor of more than 100. It is these factors which are so confusing to the public. Note that the deaths from chemical air pollution greatly exceed those from radioactive particulates.

When attempting to make a comparison between coal and nuclear fuels in the future we should not be misled by using the data on older existing plants as a standard. We must consider only the new capacity for electric power generation that will be compared with new nuclear facilities if the debate is to be meaningful. Table 17 shows the best estimates available for newer coal-fired generating stations given in the lower range of the numbers. Data for existing older plants, in urban areas, with less emission control, and burning higher-sulfur coals, are given in the higher range of numbers. The use of coals containing 3 percent sulfur is being restricted by regulation. It is not possible to place new stations in remote areas because of the line loss of electricity being delivered over long distances. However, it is also not necessary to place them in the most densely populated areas, so we have to choose an average siting that can be considered under the categories of "suburban" or "rural" as the best estimate. This

TABLE 17. Comparison of Risks in Complete Fuel Cycles for Coal and Nuclear Power Generation, Deaths per Year for Each 1,000-Megawatt-Equivalent Capacity

Coal	Deaths
Accidents and diseases in extraction (including black lung disease), transportation (mostly collisions at railroad crossings), construction, and power generation	2 to 19
Air pollution	
sulfur-related	0.07 to 295
radioactive particulates	1.9 to 3.7
Best range of the total	25 to 100
Nuclear	
Disease and accidents in mining and milling conversion, fuel fabrication, power generation, reprocessing, and waste disposal	0.05 to 1.1
Details of Pollution Hazard from Coal-Fired Generating Plants	**Deaths per Gigawatt-Year**
Sulfur-related (sulfur dioxide and suspended sulfates, causing respiratory diseases, diseases of the heart and lungs, asthmas, and chronic bronchitis	
using coal with 3 percent sulfur	6–300
using low-sulfur coal	1.2–75
using low-sulfur coal with lime scrubbers	0.12–30

Radioactive particulates (expected fatal cancers due to radioactive particulate emissions)	Existing Stations	New Stations
urban	3.7	1.9
suburban	0.25	0.015
rural	0.007	0.0044
remote	0.0014	0.00008

Source: Data from American Medical Association Council on Scientific Affairs 1978; Environmental Protection Agency 1979; Ford Foundation and Mitre Corporation 1977; National Academy of Sciences, Committee on Science and Public Policy 1979. American Medical Association data reproduced with permission from the Annual Review of the *Journal of the American Medical Association* 240, © 1978 by Annual Reviews, Inc.

Note: For a total of 430-gigawatts capacity in the U.S., 73 percent is in "new" plants, and 99 percent of the power generation is in urban areas. The weighted average death-rate per gigawatt-year due to radioactive particulates from stack emissions is 2.4.

results in the overall estimate for new installations of 25 to 100 deaths per year per 1,000-megawatt-equivalent electric generating capacity for the entire fuel cycle including mining, transportation and construction accidents, and air pollution. A generation of 1,000 megawatts (1 gigawatt) roughly provides the domestic electricity required by one million people. The deaths due to radioactive particulates in the emissions from coal plants have been discussed in more detail in the last section of chapter 8.

We turn now to a consideration of the health impact from nuclear power generation. The normal operation of a 1,000-megawatt nuclear generating plant together with its entire fuel cycle, including mining, milling, conversion, fuel fabrication, power generation, fuel reprocessing, and waste disposal, is estimated to cause between 0.05 and 1.1 deaths per year. These final totals may be compared with the hazard from coal-fired plants. Excluding major accidents the nuclear power comes out to be about 100 times less hazardous than the present day use of coal, over all. However, this comparison applies only to somatic effects. The EPA uses a simple one to one relationship between somatic and genetic effects, which roughly doubles the hazard of nuclear radiation. In addition nonfatal cancers are roughly equal to cancer deaths, and their disabling effect on the society must be considered. The future use of newly developing concepts in pollution controls in coal-fired plants will change the balance even further.

The total present electric generating capacity in the United States is about 430 gigawatts (430,000 megawatts) for coal and 55 gigawatts for nuclear. The most likely number of expected deaths in the United States per year from electric power generation is 10,000 to 40,000 (mostly from chemical air pollution) for coal, and about 20 for nuclear. It should be kept in mind that these figures are compared to 360,000 cancer deaths per year that occur naturally in the United States population, 60,000 of which are related to smoking which is a voluntary decision. These comparisons should not be used in attempting to make a decision on whether to have coal or nuclear power generation in the future.

These comparative estimates utilize risk factors that relate radiation doses in rems to ill health of the population: a procedure fraught with uncertainty because the health risks from radiation cannot be established except very approximately. A more precise comparison can be made between the radiation doses from the normal operation of seventy nuclear plants in the United States with their associated fuel cycles (including waste disposal) and our total exposure to natural and X-ray sources. Table 18 shows this comparison using the Environmental Protection Agency's figures on average radiation exposures from all sources. The risk from the nuclear generation of power thus appears to be almost trivial under normal operating conditions.

All of this discussion is highly relevant at the present time. Professor Carroll Wilson of the Massachusetts Institute of Technology has made public the final report of his international study group in the book entitled *Coal: Bridge to the Future,* in which he points out that from one-half to two-thirds of all the additional world energy needed between now and the year 2000 can be supplied by coal, mostly from the United States. Experts from sixteen different countries participated in this eighteen-month intensive study, including energy and resource leaders, business teams, and government officials, and focused their attention on the growth of coal consumption as petroleum production declines.

TABLE 18. Comparison of Radiation Exposures between Normal Background, Medical X Rays, and Nuclear Power Generation

Source	Exposure to Average Person in Population (in millirems per year)
Terrestrial: internal and external	60
Cosmic rays	30
Medical X rays, and other X rays	90
Entire national nuclear power generation	less than 1

Source: Environmental Protection Agency 1979.

Note: The nuclear power generation includes the extraction, processing of the fuels, fuel reprocessing, and waste disposal.

The highly optimistic appraisal predicts that international shipments of coal, 40 percent from the United States, should increase from ten to fifteen times its present level. With coal, the cost of energy even now is only about 60 percent that of oil, and the gap will widen in the future. At $30 per barrel for oil, the energy equivalent is $144 for a ton of coal. American coal costs only about $35 per ton to mine, and even allowing $25 per ton for the costly environmental safeguards mentioned above, the cost will be only equivalent to paying $13 for the energy equivalent of a barrel of oil. The report indicates that the coal resources of the United States are sufficient to supply the energy needs of other countries in addition to our own, and that we should replace the current major oil producing countries as an emerging energy supplier into the indefinite future. This report is held in high repute, and its conclusions have spelled out the generally existing concensus of energy experts worldwide over the last many years. However the health cost from air pollution has been the major stumbling block in the acceptance of coal. That is why it is so important to differentiate between the health hazard from existing coal-fired plants as opposed to that predicted for new installations.

Estimates of Hazards

Normal Operation of Power Reactors

There are roughly one million radiation industry workers in the United States, and literally many billions of curies of equivalent radioactivity have been produced in the atomic energy programs in the last few decades. However, the number of injuries per million workers in the industries regulated by the governmental authorities has been considerably less than one-half the rate for the rest of industry, and lately less than one-quarter of this rate. Except for the mining industry there have been no known injuries from delayed effects of radiation, although this question is difficult to assess, owing to the relatively high rate of cancers in the population as a whole from spontaneous causes.

The Atomic Energy Commission, the Energy Research and Development Administration (ERDA), and the Department of Energy (DOE), and their contractors have maintained surveyance over the exposure dosages to all of its workers. Throughout all years of employment, less than 0.2 percent of all employees received an annual dose greater than 5,000 mrem, and 95 percent received an annual dose less than 1,000 millirems, well below the prescribed limit. A study by the Federal Radiation Council, which is now under the EPA, gives an average figure of exposure for radiation workers of about 200 millirems per year. This is equivalent to less than 1 millirem per year to the population as a whole. This should be compared with the 170 millirems per year which is the average annual dose received by the population from natural background sources and X rays.

It is estimated that the increase in monitoring and control will keep the average population dose down to less than 1 millirem per year even if nuclear power developments increase several-fold in the nation in the next few decades. The Federation Radiation Council's statistics were based on a study of over 1 million person-rems of exposure of radiation workers. Using the statistical information on risk of cancer, 1 million person-rems of exposure supposedly should produce 20 cases of leukemia and about 200 radiation-induced cancers. The actual incidence of leukemia spontaneously occurring in the population is about 12,000 per year, and the death rate from all cancers in the nation is over 300,000 per year. Therefore, it can be seen that the possibility of 20 cases of leukemia and 200 cases of cancer cannot be checked statistically against this high incidence of disease occurring naturally in the population. It also shows that the entire nuclear energy industry in the nation has had a trivial effect on the nation's health so far, except for early cases in mining.

The Department of Energy semiannual report of March, 1980 (DOE-NE-0030), lists the number or prospective number of nuclear power reactors in the United States as:

Licenced	70	51,569	megawatts total
Authorized	2	910	
Being built	91	99,693	
Site authorized	4	4,112	
Planned	22	25,836	
Total	189	182,120	

About two-thirds are pressurized water reactors, and one-third are boiling water reactors. Department of Energy projections forecast an installed nuclear capacity in the year 2000 to range from 255 to 395 gigawatts (1 gigawatt equals 1,000 megawatts). However, this forecast appears to be much too high. The DOE has been directed to improve nuclear licensing, to push for completion of the Clinch River breeder reactor, and to encourage commercial reprocessing of nuclear fuel. Because the government policy has been directed toward the storage of high-level waste *after* reprocessing, it is necessary that the development of reprocessing capacity precede the development of waste repositories. It has turned out that the slowdown in the growth of nuclear power has been less due to licensing than to general factors related to the economy: the decrease in demand for electric power, the high interest rates, and the inability to construct and operate nuclear plants that are trouble free. The breeder reactor program is foundering because of the decrease in demand for nuclear fuel beyond our present uranium reserves. Private industry has been turned away from the invitation to get into reprocessing because of no guaranteed market. This leaves the waste disposal problem unsettled. Thus it seems that the government's drive toward an increased development of nuclear power has been relatively unsuccessful.

At present it is clear that the official projections on the growth of nuclear power must be revised downward to an estimated capacity of about 120,000 megawatts in the 1990s. With a present capacity of about 60,000 megawatts and slightly more than double this under construction, and no new plants being planned, it would appear that the public concern over a coming age of nuclear power should be somewhat lessened.

The release of radioactive gases into the atmosphere by 160 nuclear reactors operating or coming on stream in the 1980s is estimated (EPA 1979) to result in fatal cancers to the United States and world populations as follows:

Fatal cancers per year in the United States	
boiling water reactors	0.070
pressurized water reactors	0.094
Eventual fatal cancers worldwide due to tritium and carbon-14 emitted per year	
tritium	0.07
carbon-14	4.2

Tritium and carbon-14 affect the world population because of their long half-lives. The conversion factors used for the number of worldwide fatal cancers per curie released to the atmosphere are 6.0×10^{-7} for tritium, and 4.1×10^{-3} for carbon-14.

Uranium-238 is the parent of the uranium series and is present in the amount of 99.28 percent in ordinary uranium. Its next most abundant isotope, uranium-235, is the parent of the actinium series, and is present in the amount of 0.7 percent in common uranium. Because it fissions readily in the presence of neutrons, uranium-235 is the principal substance used in the fission process of generating nuclear power. Plutonium-239 is produced in a uranium reactor, and is, therefore, man-made. It will also fission in the presence of neutrons and is useful in nuclear weapons.

Enrichment of uranium in the isotope uranium-235 is carried out in gaseous diffusion plants in which a process takes place using gaseous uranium hexafluoride. There are very large plants at Oak Ridge, Tennessee; Paducah, Kentucky; and Portsmouth, Ohio. Fuel element fabrication of the natural or enriched uranium then shapes it to the particular needs of its end use. Reactors using these fuel elements may be for power generation or the production of plutonium. In power generation the reactors may be used for the generation of electricity for the civilian population, or for the navy, or for research.

Risks associated with fuel enrichment plants are limited to the workers in the plants almost entirely. The decay products of uranium such as radium and radon have been previously removed from the uranium when it arrives at the plant. No statement can be made specifically regarding the risk to the workers due to inhalation of gaseous uranium compounds used in the enrichment process, except the overall safety record of the AEC, ERDA, and DOE. These agencies have continuously maintained a monitoring of the ill health and accident rate of the sizable population of workers under their jurisdiction, and have a solid record of achievement showing less than one-half the rate of all other industries.

Risks from Fuel Reprocessing

Spent fuel, in which only a small proportion of uranium is generally consumed, may be reprocessed in plants in which the fuel elements are dissolved and the unused uranium and plutonium are recovered. Unwanted fission products then have to be disposed of in some form of waste storage, sometimes after the removal of certain fission products useful in research or other applications. The plutonium may be stockpiled for consumption in reactors, or used in the weapons inventory.

A possible shift from the uranium oxide fuels in light-water reactors at present to plutonium fuels in fast breeder reactors by the year 2000 might result in the need to reprocess about 17,000 tons of plutonium for fuel and 3,000 tons of uranium fuel per year. The EPA has estimated an almost exactly similar average dose to the United States population from fuel reprocessing at that time as given above for power reactors. This is a whole body dose from external gammas of 0.2 millirems per year, with an additional dose of 0.04 millirems per year from tritium, 0.04 millirems per year whole-body from inhaled krypton-85, and 1.6 millirems per year to the skin from krypton-85 in the atmosphere. As defined earlier, this is well below any detection level for the ill health of the population. Plutonium is an extremely dangerous material if it finds its way into the human lungs. The controversy over plutonium and its possible escape as an atmospheric contaminant is part of the discussion related to breeder reactors, the use of which in this country has so far been denied by the United States government. This subject is beyond the scope of this book. Plutonium release in a vaporization of the core of a water-cooled reactor is included later in the section on the estimate of risk associated with a catastrophic accident.

Waste Disposal

There is no major problem associated with the disposal of so-called low-level radioactive wastes. They come from hospitals, research laboratories, and industry and can be diluted and dispersed at levels approaching the natural background. Part of the problem is a public reaction to any radioactivity with a natural suspicion that the agencies in charge of the operation may not be candid in all facets of control and monitoring. The problem is, therefore, political rather than real for the most part and will have to be resolved by the political process.

Of much greater concern to the public is the disposal of the so-called high-level radioactive wastes. These come from the treatment of spent fuel elements from nuclear reactors, in fuel reprocessing plants. Nongaseous radioactive elements are taken out of solution and incapsulated in one manner or another so that they can be disposed of in solid form without the large volume of water in which they were formerly dissolved. One of the most promising processes of incapsulation is to form the waste material into glass blocks by melting them into a matrix material.

In order to visualize the amount of solid waste per year that would have to be disposed of by the year 2000 one can visualize a cube about 25 feet on a side. These are not very large volumes, but the material cannot be stored in close-packed form because heat must be permitted to escape. The total heat generated by twenty years of accumulated waste would be about 1,000 megawatts-equivalent, mostly from radioactive strontium and cesium. This heat may be recovered in the future. Present plans for disposal include storage underground either in large bodies of rock or in abandoned salt mines. It is probable that the waste disposal sites would be constantly controlled and monitored for several decades, until future experience indicates that they may be safely left in isolated repositories.

In November, 1979, the National Academy of Sciences held a forum entitled "Nuclear Waste: What to Do with It?" The proceedings of this forum have been published by the academy. The eminent members of the panel represented the fields of chemistry, geochemistry, biochemistry, nuclear engineering, economics, and politics. The panelists generally agreed that the level of radioactive contamination of the environment resulting from any waste disposal procedures should be no more than the variations in the natural background that the population is exposed to in its normal travels about the country. This includes airborne and waterborne contaminants.

The two radionuclides that are most hazardous during the first few hundred years are strontium-90 and cesium-137. It was agreed that these two must be kept out of the biosphere as completely as possible for a length of time that was specified as being between 500 and 1,000 years. By that time these radioactive isotopes would have decayed enough so that they would no longer be a problem. After that time the radioactivity of the waste is mostly in the heavy actinide elements and their decay products. These elements have long half-lives so that they will continue to be radioactive into the long-term future. However their levels of radioactivity, after about 1,000 years, would be not much greater than the radioactive material in a medium-grade uranium ore deposit. Therefore the disposal of the waste products in the repository may be considered as being only important for the first 500 to 1,000 years, after which the problem is no greater than the original presence of the uranium ores below the ground.

Let us now consider some of the geological and geochemical facts related to the burial of radioactive glasses, if the wastes are going to be stored below ground in this form. An important mineral used for age dating in geology is a zirconium silicate named zircon. This mineral occurs in tiny grains in granitic rocks and is separated from the rock in order to determine the age of the rock by radioactivity methods. Its behavior has therefore been well studied. Uranium and thorium also occur in minor amounts in the zircon so that the mineral is radioactive and contains in addition all of the decay products of the uranium and thorium series. The radioactivity damages the crystal structure so that each zircon grain gradually becomes glassy, or metamict, losing its crystalline nature as seen under X-ray examination. Most old zircons are therefore a zirconium silicate glass as a result of the alpha particle bombardment. Nevertheless they retain their original shape over periods of time that have been measured as far back as four billion years, almost the age of the earth, and at temperatures of several hundred degrees. Sometimes the glassy zircons take on water by hydration, but still retain their shape and do not dissolve or disintegrate. They lose some of their decay products, partly by the release of radon gas, but essentially none of the uranium and thorium, which are like the heavy actinides that would be in the silicate glass of the waste disposal process.

In addition to observing the stability of zirconium silicate glass over thousands of millions of years, the geologist also observes the stability of obsidians, or black volcanic glasses, of the kind used by southwestern Indians for arrow tips. They may be tens of millions of years in age and yet show essentially no change except a very thin rind (less than 1 millimeter) of hydrated material caused by being soaked in water during their lifetime. The thickness of the hydration layer is used to determine the age of the glass, so that it is clear that the glass is not going into solution to any extent, even when hydrated. After millions of years the glass may start to recrystallize. Thus if we have radioactive elements in a man-made glass we will have the radioactivity working toward keeping the glass metamict, or glassy, as in the case of zircon, as opposed to a natural devitrification process in which the glass tends to crystallize. These two processes work against each other, especially if the glass is raised in temperature. In short, the glass would not tend to disintegrate, go into solution, or lose its heavy actinide constituents even under warm or wet conditions.

The geochemical behavior of elements diffusing through a rock that is saturated with water is well known. The water in the capillary cracks of the rock is saturated in all of the chemical components of the rock, and if there is a source of new materials, such as leaking from a waste disposal unit, these materials would be precipitated as minerals and form part of the rock. For example, strontium would be precipitated in calcium-bearing minerals, such as feldspar or apatite; cesium would be precipitated in potassium-bearing minerals, such as mica, and so on. In other words, these so-called contaminants would form an envelope around the waste disposal unit as part of the rock itself. If an open channel way conducted water through the system at a flowing rate, these elements would be carried by the water and not be precipitated. The choice of the disposal site, therefore, has to ensure that there will be no flowing water in an open fissure. But it does not have to ensure that there will be no stagnant water or water that is very slowly permeating through the rock.

Therefore, one can visualize radioactive waste made up of silicate glass in canisters being deposited in a bore hole in a large body of granite or basalt at a concentration level that will not permit the waste to heat up above a certain temperature. Then the bore hole subsequently would be filled to its top by an impermeable substance so that the waste unit is left sealed below some considerable depth of rock. Surrounding this bore hole would be other holes drilled to monitor any possible migration of radioactive materials away from the disposal unit, even if none were expected to leave the glass at all. An alternate choice would be a repository in a large underground body of salt which is known to have existed for millions of years without any surface water penetration.

One can reasonably ask about the possibility of some later civilization becoming exposed to one of these repositories by accidentally excavating the rock down to that depth. One answer that can be given is that the chance of a later civilization digging a shaft several hundred meters down into a blank body of granite located somewhere in the continent would be extremely remote. Even if this occurred, the radioactive hazard would not be much greater than that faced by ancient miners, unaware of radioactivity, entering a rich uranium vein in search of silver or other metal. This, in fact, actually happened.

The panelists in the National Academy forum mentioned earlier gave the impression that the problem of waste disposal can be taken care of with reasonable assurance to the public that contaminants reaching the biosphere will be less than the normal variations of the background radiation. Delay in choosing waste disposal sites appears to be due to the time it takes to decide and make tests on the best out of many choices. One of the problems of waste disposal appears to be the public concern itself, which appears in the form of refusal by individual states to accept the siting of a repository. In the meantime it is believed by the panelists that there is no great hurry to dispose of the waste because the present-day temporary storage procedures are now adequate and safe. This does not mean to say that there have not been poorly executed or devised depositions of waste in the past, about which the public had a right to be concerned.

Estimates also state that even the most expensive methods of waste disposal will make only a minor variation in the cost of the final electricity. Certainly there will be serious occupational risks unless the handling facilities are expensive, well-designed, and function properly. The process of glassifying the waste is well developed. It is being done routinely in France. The fuel elements may be taken from a reactor, stored in water pools, and then either reprocessed or encapsulated, and finally disposed of either in the form of glassified wastes or encapsulated fuel elements in the final repository. Each nuclear plant produces about ten canisters of waste per year. The present state of the art involves experimental and demonstration disposal sites already being used in which the buried canisters will be retrievable for several decades during the long period of testing.

Catastrophic Nuclear Accident

We come now to a very brief consideration of the possibility of catastrophic nuclear accident. This has been the cause of more public concern than any other aspect of the nuclear industry, and it is the most difficult of all of the risks to assess. For a light water thermal power reactor, the major catastrophic event that could result in the release of large quantities of radioactive materials to the environment would be the loss of coolant accident. Although the absence of water moderator would stop the fission process, the decay heat due to the inventory of radioactive material could result in a meltdown of the reactor core, and a breakthrough of the containment. Volatile radioactive elements escaping would be the noble gases, iodine and tritium. Nonvolatile elements include all fission products, many activation products, and uranium and plutonium. A release in the form of a radioactive cloud moving downwind could expose a large number of people.

In the Reactor Safety Study (RSS) Report to the Nuclear Regulatory Commission in 1975 from the Rasmussen Committee, a breach of containment and release into a heavily populated area of 3,200 square miles was estimated to result in 3,000 early fatalities, 45,000 radiation sickness victims, 45,000 latent cancer deaths, 240,000 nonfatal thyroid nodules, and 30,000 genetic disorders. It was further estimated that such an event would occur once in 200 million reactor-years. The Ford Foundation study found this estimate to be too low by a factor of 500, giving a value of one such accident in 400,000 reactor-years. This would mean one such event happening every 4,000 years if there were 100 reactors.

Taking the figure of 100,000 deaths (from somatic and genetic diseases) we find a risk of about 10 deaths per gigawatt-year that should be added to the nuclear side of the account in table 18. However, can we honestly attempt to divide such numbers? If the nuclear industry were to continue for 4,000 years, possibly yes. At the present time there is little interest in the United States in favor of breeder reactors and, as will be seen in the next section, the supply of domestic uranium will only permit a limit of about 400 nonbreeder reactors in all, ever to be installed in the United States. This would appear to reduce the chance of a single catastrophic accident to a low level of risk, but the averaging of such a death toll is meaningless.

These estimates were made before the Three Mile Island (TMI) incident. Since TMI there has been a general overhaul of all reactor safety regulations, equipment, and procedures. An interesting improvement has been the actions taken by state governments toward meeting evacuation and ingestion control plans specified by the Federal Emergency Management Agency and the Nuclear Regulatory Commission. The plans call for detailed studies of evacuation areas, routes, and destinations, as well as official state and federal preparedness to supply assistance toward evacuation in a matter of hours following an accident. Evacuation areas are specified as 10 miles in a radius surrounding a power reactor. In addition to the evacuation preparedness the guidelines call for a 50-mile radius to be monitored as an ingestion area, in which possible nuclear contamination might be ingested by dairy cattle. The testing will be expanded to include watersheds, and all foodstuffs coming from the area. The TMI accident was fortunately contained. The maximum dose of radiation received by any person downwind from Three Mile Island was only about 100 millirads.

Limits to the Magnitude of the Nuclear Industry without Breeder Reactors

All electric generating power plants operate at some level below their theoretical installed capacity. The ratio of actual power generation to maximum theoretical generation is known as the capacity factor. For example, the capacity factor for new coal-fired power stations is estimated to be about 65 percent, so that in all calculations of the use of coal or uranium one must be careful to make allowances for this. In considering a new model of coal-fired power plant typical of those that will be installed in the future, the EPA estimates that the actual generation of 1,000 megawatts will require 2.7 million metric tons of coal per year.

This may be compared with the use of uranium in nuclear power plants. There is considerable disagreement on the amount of electric energy that can be obtained from uranium because of a number of factors entering the calculation. These include the loss of uranium-235, and the percentage enrichment in the enrichment plant, the probability of uranium-235 fission and burnup of uranium-235, fission other than uranium-235, or the number of supplemental fissions per fission of uranium-235. Estimates vary between 20 and 50 million kilowatt-hours per short ton of uranium oxide. The Energy Research and Development Agency has estimated 32. Using this figure we see that 1 short ton of natural uranium, not the oxide, would produce 38 million kilowatt-hours of electricity, and that 1 ton of uranium is equivalent to 13,000 tons of coal in actual electricity produced.

In comparing uranium ores containing 1,500 parts per million of uranium we must allow for the recovery of the uranium from the ore and of the uranium-235 from the total uranium derived from the ore as it goes through the enrichment process. If we take this total loss to be about 50 percent we find that 1 ton of present-grade uranium ore is only equivalent to about 10 tons of coal in its final yield of electric energy. If we further consider the cost and general difficulty of obtaining a ton of uranium ore compared to a ton of coal we see that the total cost, in both dollars and energy, of exploration, mining, milling, uranium extraction, and enrichment is large compared to the preparation of coal ready for its burning. Not only do the uranium deposits have to be discovered, by costly prospecting, but their occurrence underground means much development work in the form of exploratory and development drilling, assaying, and other preparatory work. Also each ton of ore requires the removal of many tons of overburden if the mine is to be operated from the surface, or many feet of underground workings if it is to be operated underground. Compared to the large-scale mining of near-surface coal, the energy advantage disappears rapidly. It turns out that the present ore grades are almost as low as will be permissible on the basis of a fuel-cost comparison.

Using this consideration we can now go to estimates of uranium ore reserves. Using a base price of $30 per pound for uranium in 1977 and the best estimates of ore reserves at that time, we can extrapolate into the future on the basis that increased costs will be balanced approximately by increased price of uranium, leaving the cut-off grade of ore about the same. There have been several recent studies attempting to assess the domestic uranium resource base. The subject is complicated by the fact that ore reserves are broken down into categories that depend on the degree of speculation regarding the deposits that are not yet discovered. It is common practice to divide the total reserve picture into three parts: proven reserves, already measured in place; potential resources made up of probable ore that represents reasonable continuations of single ore bodies or districts; and speculative potential resources made up of totally unknown deposits that are expected to be found as exploration continues. In addition, all of these resources have to be based on some projected cut-off grade, which is the grade of ore below which it is uneconomical to mine at that particular time in the future. As we have seen above, this cut-off grade cannot be much less than present grades, unless the industry goes to breeder reactors.

Central to all actual estimates is the data base developed by the Grand Junction Office of the Department of Energy. This organization carries out a continuous monitoring of exploration and discovery information by all groups engaged in the business, using congressional authority. Although proprietary data are kept confidential, they appear in final summations that are made available to the public periodically. It is not possible for any individual outside of this office to obtain similar data by any other means. The Grand Junction Office delivers reserve information to the Department of Energy, and to various groups, who further analyze the resource picture. Differences in the resource estimate, therefore, appear only as a result of treating the probable and speculative categories in different ways. Of the more recent studies we have the 1977 report of the Nuclear Energy Policy Study Group, sponsored and administered by the Ford Foundation and the Mitre Corporation, and the 1978 report released by the subpanel of the NAS-NRC Committee on Nuclear and Alternative Energy Strategies (CONAES), giving estimates of the domestic resources and projections of discovery rates. The first is a more optimistic report, whereas the second leans toward the conservative side. The Ford-Mitre study claims that the resource base is adequate, while the CONAES study presents arguments that the Grand Junction Office has overestimated the ore potential. Selected estimates are summarized in table 19. We may use the figures of 640,000 tons of uranium oxide as the proven reserve, and an additional 1 to 3 million tons as possible future supplies.

From the information given earlier we see that a 1,000 megawatt nuclear plant operating at 80 percent capacity uses about 167 tons of natural uranium per year, or 6,680 tons

TABLE 19. Domestic Uranium Resource Estimates, 1976

	Reserves[a]	By-products	Potential Resources	Total
Best estimate	640,000 (ERDA)	60,000	1,060,000	1,760,000
Lower-limit estimate	480,000	20,000	500,000	1,000,000
Higher-limit estimate	640,000	140,000 (ERDA)	3,000,000 (ERDA)	3,780,000 (ERDA)

Note: ERDA is the Energy Research and Development Agency that is now under the Department of Energy.

a. Tons of U_3O_8 at the $30 per pound U_3O_8 cut-off cost during the total forty-year life of the plant. The known reserve of 640,000 tons would then fuel a total of 96 plants, each producing 800 megawatts, for their entire life span. Using the optimistic and pessimistic estimates of additional possible uranium supply we can see only an additional 150 to 450 plants in all the time span of nuclear power, unless breeder reactors are used. If there is a decision to go to breeder reactors to extend the fuel supply, it will be thirty years or more before any substantial addition of nuclear fuel could be available.

At the present time there are about 50,000 megawatts of actual nuclear power generation, with about 100,000 megawatts of additional capacity on order. This is getting close to the limit of ore supply, in the near term at least. It is not unlikely that many of these orders will be dropped as a result of extraordinary increases in cost of construction, due largely to delays in obtaining permits. For example the two Seabrook nuclear units in New Hampshire will have a construction cost estimated to be in excess of $1.70 per watt, from reports available. With a planned capability of 2,300 megawatts this will mean a total construction cost of about $4 billion. The total debt service over the thirty years of operation will add up to $10 billion if tax-free interest rates remain at normal levels. Despite these excessive costs the fuel charge in the average individual's monthly bill, if fossil fuels were used instead, would still exceed the debt service and fuel charge combined, when nuclear fuel is used.

10

Summary of Technical Information

A comparison of per capita radiation doses from all sources in the next decade is presented in table 20. Most of the figures have been reassessed by the Environmental Protection Agency, and are for the average person living in the most typical radiation environment, using the average rate of X-ray exposure for diagnostic purposes per year, and are under the assumption that there has been no major accident or act of war that involves a release of radioactive materials. As seen from the table, the natural terrestrial and cosmic radiation backgrounds, together with the normal use of X rays, make up about 98 percent of the dosage of radiation received by the population. Additions from the mining, milling, enrichment, processing, burning, reprocessing, and disposal of nuclear materials and their waste products both in weapons and power generation, make up less than 1 percent. This is the principle message of this book.

Also given in table 20 is a genetically significant dose, which is less than the whole-body dose, representing almost solely the cumulative dose to the gonads in the preparental period of zero to thirty years, for the younger part of the population. These figures are used to estimate the rate of incidence of harmful genetic effects in the next generation and in the subsequent population at equilibrium. The number of genetic diseases and abnormalities resulting from this average whole-body exposure of 170 millirems per year, or 5,000

TABLE 20. Estimated Radiation Exposures, United States Population Average, in Millirems

Radiation	Whole-Body Dose		Genetically Significant Dose	
	Per Year	60-year Lifetime	Per year	Preparental, 30 years
Natural				
Terrestrial, external	35	2,100		
Terrestrial, internal, exclusive of radon[a]	28	1,680		
Cosmic	43	2,580		
Total	106	6,380	90	2,700
Man-made				
Medical X Rays	79	4,740	45	1,350
Radiopharmaceuticals	13	780		
Fallout	4	120		
Nuclear power	0.1	6		
Tritium in atmosphere	0.03			
Krypton-85 in atmosphere	0.001			
Other (mostly burning of fossil fuels)	2	120		
Radon and its daughters, from all sources	[a]	[a]		
Total	98	5,766	45	1,350
Total	204	12,146	135	4,050

Source: Some data from Environmental Protection Agency 1979.

a. Normal atmospheric radon and its daughters will cause an average annual dose of 250 millirems, or a lifetime dose of 15 rems, to the tissues of the lungs. Man-made additions may double this in some areas and in some building interiors.

millirems in the preparental age range, is predicted to be 25 to 325 in the first generation, and 300 to 5,500 in future generations at equilibrium, for each 1 million persons in the population. In a future population of 300 million persons in the United States the total number of persons who will suffer a genetic problem during their lifetime will, therefore, be 90,000 to 1,650,000, or the number of new cases a year will be 1,200 to 22,000, due to average levels of radiation. This average lifetime dose of whole-body radiation is seen in the table to be about 11,000 millirems. Remember, more than 99 percent of this is *not* due to the advent of nuclear power.

The table also gives a comparison between the radiation effects of electric power generation by nuclear fuels versus coal. A comparison of the entire fuel cycles for these two energy sources, based on plants of equal capacity, indicates that coal-fired plants have a greater impact on public health than nuclear plants, even from nuclear radiation.

In table 21 a list of annual whole-body doses is given for

TABLE 21. Ranges of Radiation Exposure for Specific Groups, or Subsets of the Population, or Individuals

Radiation	Whole-Body Doses (in millirems per year)
Natural	
Terrestrial, external, terrain with higher levels of activity	100 to 500
Terrestrial, external, uranium miners	5,000 maximum
Terrestrial, internal, uranium miners (specific organs)	15,000 maximum
Cosmic, higher elevations, on land surface	100
Cosmic, jet flights (airline personnel)	up to 500
Man-made	
Medical X rays (local exposures, not whole body, per film)	
head and neck (CT scanner)	1,000
chest (radiographic)	30
abdomen	up to 5,000
teeth (mandibular bone and marrow)	1,000
gallbladder	50
Worker in a radiation industry	5,000 maximum

Note: Upper limits set by federal regulation

subsets of the population who are exposed to abnormal levels of radiation. In the previous tables these were included but diluted by the remainder of the population to such an extent that they do not show more than low average values. The lifetime exposures of these subsets may reach levels which are observably affected in increased morbidity and mortality in local settlements or in groups of individuals. Generally the increased risk is compensated for by some benefit that is recognized and accepted by this group. It can be seen from this table that the average lifetime exposure of 11,000 millirems found in table 20 can easily be increased by a factor of two or three in subsets of population who are exposed to higher levels of natural or man-made radiation, with a consequent doubling or trebling of the predicted harmful genetic effects in their descendents.

A summary of estimated cancer deaths from average natural and X-ray radiation exposures is given in table 22 using

TABLE 22. Summary of Estimated Cancer Deaths in the United States Due to Average Radiation Exposure from Natural and Man-made Sources

Source	Predicted Fatal Cancers per year
Terrestrial	
External	1,200
Internal	1,000
Cosmic Rays	1,400
Technologically Enhanced	
Natural Radiation	
Building interiors	700
Radioactive particulates from coal-fired power plants	200
Soil tillage	30
Uranium mining and milling	15
Phosphate mining and processing	5
Ground water treatment	5
Man-made Radioactivity and Radiations	
Medical X rays	3,000
Nuclear fuel cycle	4–80
Catastrophic nuclear accidents	(0)

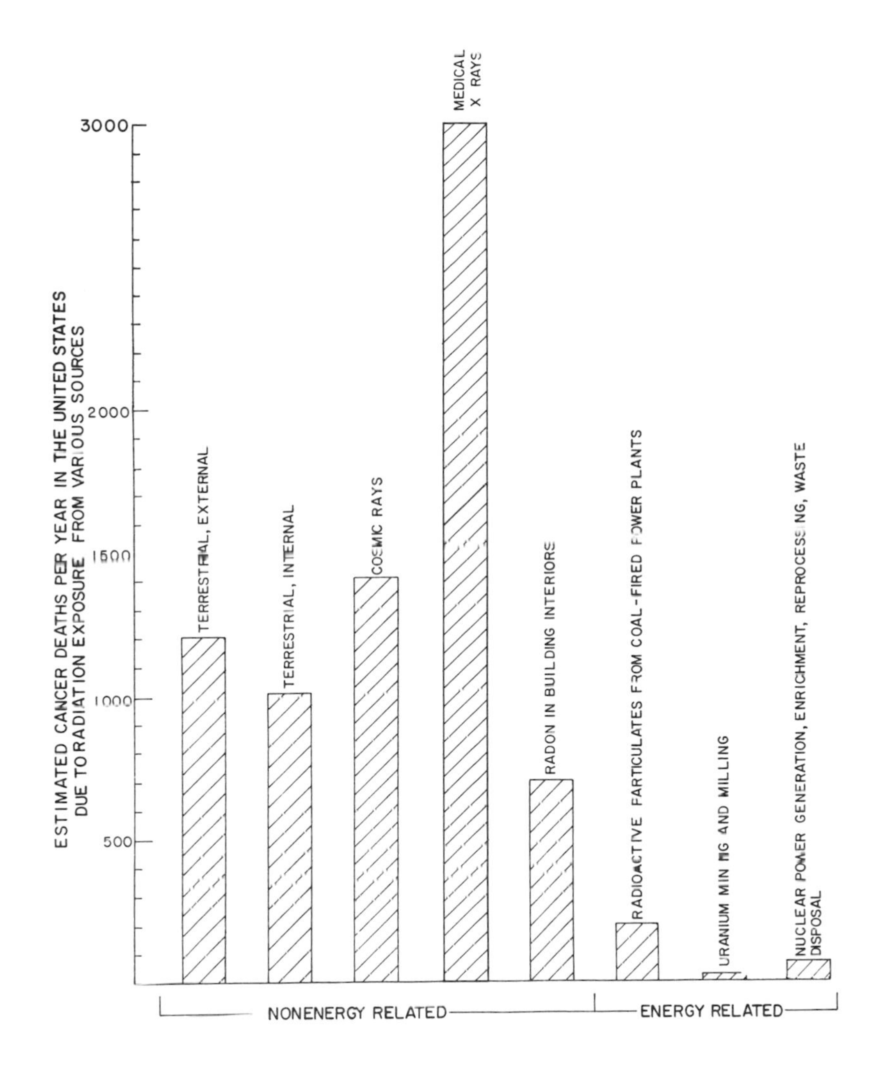

Fig. 13. Comparison of the health hazards from various sources of radiation, showing the relatively small proportion coming from the generation of electric power and the much larger proportion from nonenergy related sources that cannot be reduced, and which have always been present.

the EPA estimate of 1 in 5,000 for cancer fatalities resulting from 1,000 millirads of exposure. The total of about 5,000 per year can be added to the 1,200 to 22,000 radiation-related genetic diseases and abnormalities that will appear each year in the future population of the United States. The overall effect of this radiation is also estimated to account for 1 to 5 percent of the general ill health of the population. On the other hand, the total annual United States death toll resulting from the use of nuclear energy, assuming that the chance of a catastrophic nuclear accident is essentially zero for the total time span of using up the available uranium, is estimated to be between 20 and 95 (see fig. 13). The understanding that this relatively small additional risk results from the use of nuclear energy for peaceful purposes provides the basis for this book: learning to live with nuclear radiation.

Epilogue

It is a commonly accepted scientific estimate that of the millions of advanced civilizations on the habitable planets of our galaxy, a sizable fraction will have destroyed themselves following the discovery of nuclear energy. However this is the nature of evolution—it is inefficient and wasteful. The civilization on Planet Earth thus faces a difficult period in its evolving level of advancement. We can come through it, or we can fail. By sheer determination we can, and must, learn to live with the enormously dangerous presence of nuclear energy in the implements of war. This will take all of the intelligence and pressure that mankind can muster. Because we are not yet wise, we will survive this period by a process of trial and error, muddling through, noble and ignoble acts, and bitter setbacks—but always determination. This must be our goal.

We cannot turn back the clock or hope that the knowledge of nuclear energy can be forgotten. A single nation armed with deliverable nuclear bombs could command all others to obey. It therefore seems that the concept of nuclear deterrence by the threat of retaliation is here to stay. What, therefore, can we do?

How much of a nuclear deterrence is needed by any nation to prevent itself from being attacked? It would seem that it is not the number of warheads, but equality of offensive and defensive capability. The arsenals of the super powers could be reduced if some way could be agreed upon to achieve equality with less. At present it seems that sheer exhaustion of resources in the form of military spending brings about the terminal degree of preparedness. It is in an area such as this that public pressure can be applied.

Large segments of the scientific community have devoted themselves to an assiduous search for a solution to the complex and crucial problems of the arms race. The Pugwash Conferences attended by scientists from opposing nations, the constant speaking out through the Bulletin of the Atomic Scientists, convocations throughout colleges and universities nationwide, and the lobbying of the senate by groups such as the Council for a Livable World demonstrate the serious attempts being made. There are outcries from other countries. Europe is in a turmoil of resistance against the deployment of short-range nuclear missiles by the NATO forces, put there to counterbalance similar missiles on the Soviet side. These activities are not trivial. They are having their effect, and sooner or later, together with new voices raised by all the peoples of the world, there will be a turning point.

It is possible that the policing of world law and order could occur as a result of a stable stalemate of nuclear counterbalance between super powers. Any nuclear exchange or conflict between minor nations would be too great a destabilizing influence to be tolerated. This is to some extent the situation we have today. It is possible that an interim period of such tenuous peace could bring us through to some enlightened age in which the balance of terror will subside, and the policing action only will remain.

Turning to the question of peaceful applications of nuclear energy we can visualize optimal programs which would be a blessing to mankind. There is no doubt of this. However, it is the less than optimal implementation of such programs that may offset the gains. It is thus the duty of citizens at large to insist that any use of nuclear energy should be bound by regulations that are encompassing enough to ensure that the gains outweigh the losses. It is not the duty of citizens to attack blindly, to obliterate attempts to achieve workable utilizations, on the arbitrary basis that anything related to nuclear energy is bad. Certainly the decisions should be made by an informed public, rather than one which has reacted only as a result of propaganda.

We have seen that as a result of public pressure the cost of new nuclear power installations has been so increased as to have slowed the growth of the industry severely. Protests have added to the conservatism of the Nuclear Regulatory Commission, the Environmental Protection Agency, and state authorities from which approvals have to be obtained. The federal regulatory bodies may grant an approval, but these must be upheld by the courts on appeal by opposition groups. The result is that there are such costly delays, and such uncertainty in the final approval, that many new installations in the planning or even the construction stage have had to be abandoned. There is no question but that these activities have been successful. On the other hand the nuclear generating plants are not being replaced by benign energy sources, but by coal. The annual loss of life stemming from the burning of coal is now as great as is estimated for a major accident in a nuclear reactor involving meltdown and breach of containment: an exceedingly unlikely happenstance.

We are clearly in a period of trial and error. Government mandates have forced a conversion of oil-burning plants to coal, gas, or nuclear. No other energy source is yet ready to take their place. Conservation may have slowed the demand for electricity, but in coastal areas where oil has been used most there is also a concentration of industry. What, therefore, should an electricity utility company do to replace the loss of oil-fired, and dangerously obsolete coal-fired, capacity?

What is happening is that both fossil and nuclear-fueled plants are continuing to be tested. Both need to clean up their acts. Waste disposal methods and sites are still waiting to be selected. New methods of reducing the contaminants in the effluents from coal-fired plants are still being developed. The costs, delays, and problems of financing of new nuclear installations are tilting the balance away from nuclear power at present, in the decisions of utility companies. On the other hand, polls of public opinion seem to be favoring nuclear power over coal. It is clear that this uncertainty will resolve itself as the testing period proceeds. Perhaps this is the best way.

While still considering only peaceful uses of nuclear energy, if a person is neither strongly pro- nor antinuclear, what should he or she do? The answer, according to the presentation in this book, might be to relax a little. Become enlightened on the subject as much as possible; vote with knowledge rather than ignorance; be pro- or antinuclear, as the case may be, but be not greatly alarmed. The radiation hazard from nuclear power is so much less than from natural radioactivity and medical X rays that there should be no great component of fear in one's decisionmaking.

On the other hand, in regard to the military use of nuclear energy for destructive purposes, the reverse is true. Be afraid; be active; the greatest challenge to life on this planet is at hand. We must make it through! We must learn to live with nuclear energy, not perish by misusing it.

Glossary

AEC

Atomic Energy Commission (discontinued with formation of ERDA and NRC on January 19, 1975)

alpha particle (α)

A positively charged particle made up of two neutrons and two protons, and thus identical to the nucleus of the helium atom. It is radioactively emitted in the alpha decay of plutonium and of several of the isotopes in the decay series of uranium and thorium.

background radiation

Radiation in the natural human environment originating from cosmic rays and from the naturally radioactive elements of the earth, including those within the human body

beta particle (β)

An elementary particle emitted from a nucleus during radioactive decay. It has a single negative electric charge and a mass equal to 1/1837 that of a proton. A beta particle is identical to an electron.

boiling water reactor (BWR)

A type of nuclear power reactor that employs ordinary water (H_2O) as coolant and moderator and allows bulk boiling in the core so that steam is generated in the primary reactor vessel

Ci

Curies

cm

Centimeter

Curie

The basic unit used to describe the intensity of radioactivity in a sample of material. One curie (Ci) equals 37 billion disintegrations per second, approximately that occurring in 1 gram of radium.

decay
The spontaneous radioactive transformation of one nuclide into a different nuclide or into a different energy state of the same nuclide

decay chain
The sequence of radioactive disintegrations in succession from one nuclide to another until a stable daughter is reached

DNA
Deoxyribonucleic acid. Molecular strands with encoded genetic information

DOE
Department of Energy. Established by Executive Order in October, 1977. Comprises the following former agencies: Energy Research and Development Administration, Federal Energy Administration, Federal Power Commission, and parts of the Department of Interior

dose
The energy imparted to matter by ionizing radiation per unit mass of irradiated material at a specific location. The unit of absorbed dose is the rad. A general term indicating the amount of energy absorbed from incident radiation by a specified mass

dose commitment
The integrated dose which results from an intake of radioactive material when the dose is evaluated from the beginning of intake to a later time (usually fifty years). It is also used for the long term integrated dose to which people are considered committed because radioactive material has been released to the environment.

enriched uranium
Uranium in which the percentage of the fissionable isotope uranium-235 has been increased above the 0.7 percent contained in natural uranium

EPA
Environmental Protection Agency

ERDA
Energy Research and Development Administration (the nuclear program components of ERDA were formerly part of the AEC), now part of Department of Energy

fission
The splitting of a heavy nucleus into two roughly equal parts (which are nuclei of lighter elements), accompanied by the re-

lease of a relatively large amount of energy and frequently one or more neutrons

fission products
Nuclei formed by the fission of heavy elements. Many are radioactive, e.g., strontium-90, cesium-137.

fuel cycle (nuclear, reactor)
The series of steps involved in supplying fuel for nuclear power reactors. It includes mining, refining, the original fabrication of fuel elements, their use in a reactor, chemical processing to recover the fissionable material remaining in the spent fuel, reenrichment of the fuel material, and refabrication into new fuel elements.

fuel separation (fuel reprocessing)
Processing of irradiated (spent) nuclear reactor fuel to recover useful materials as separate products, usually involving separation into plutonium, uranium, and fission products

g
grams

gamma rays (γ)
High-energy, short-wavelength electromagnetic radiation emitted by a nucleus. Gamma radiation usually accompanies alpha and beta emissions and always accompanies fission.

gastrointestinal dose (G.I. dose)
The dose to the stomach and lower tract of humans and animals via external exposure or via internal transport of radioactive material

gigawatt
10^9 watts

gigawatt-hour
One million kilowatt-hours

half-life
The time in which half the atoms of a given quantity of a particular radioactive substance disintegrate to another nuclear form

isotope
One of two or more forms of an element that differ in atomic weight. Nuclides with the same atomic number (i.e., the same chemical element, characterized by the number of protons contained in the atomic nucleus), but with different atomic masses (i.e., different numbers of neutrons contained in the nucleus). Although chemical properties are the same, radioactive and nu-

clear (radioactive decay) properties may be quite different for each isotope of an element.

LET

Linear energy transfer. The physical measure that specifies the relative effectiveness of equal absorbed doses of different particles or rays in damage to the body. See *rem*

light water

Normal water (H_2O), as distinguished from heavy water (D_2O)

light water reactor

A reactor in which ordinary water (H_2O) is used as the coolant and moderator. In such reactors the water is either allowed to boil (boiling water reactor or BWR) or is pressurized to prevent boiling (pressurized water reactor or PWR).

long-lived nuclides

Radioactive isotopes with half-lives greater that about thirty years. Most long-lived nuclides of interest to waste management have half-lives on the order of thousands to millions of years.

low-level waste

Wastes containing types and concentrations of radioactivity such that shielding to prevent personnel exposure is not required

m

meter; as a prefix, milli-

MeV

Million electron volts

micro (μ)

Prefix indicating one-millionth (1 microgram = 1/1,000,000 of a gram or 10^{-6} gram)

milli-

Prefix indicating one-thousandth

millirem

One-thousandth of a rem

ml

milliliters

mrem

millirem

nano

Prefix indicating one-thousandth of a micro unit (1 nanocurie = 1/1000 of a microcurie or 10^{-9} curie)

natural (normal) uranium

Uranium as found in nature. It is a mixture of the fertile uranium-238 isotope (99.3 percent), the fissionable uranium-235 isotope (0.7 percent), and a minute percentage of uranium-234.

neutron

An uncharged elementary particle with a mass nearly equaled by that of the proton. Neutrons are part of the fission chain reaction in a nuclear reactor. They can also be generated by spontaneous fission and by collision of high energy gamma rays and alpha particles with some nuclei.

NRC

Nuclear Regulatory Commission (formerly part of AEC)

nuclide

Any atomic nucleus specified by its atomic weight, atomic number, and energy state. A radionuclide is a radioactive nuclide.

nuclear radiation

Particles and electromagnetic energy given off by transformations occurring in the nucleus of an atom

pCi

picocuries

person-rem

Used as a unit of population dose obtained by multiplying the average dose per individual expressed in rems times the population affected

pico-

Prefix indicating one-millionth of a micro unit (1 picocurie = 1/1,000,000 of a microcurie or 10^{-12} curie)

plutonium

A radioactive element with atomic number 94. Its most important isotope is fissionable plutonium-239, produced by neutron irradiation of uranium-238.

population dose (population exposure)

The summation of individual radiation doses received by those exposed to the source or event being considered

ppm

Parts per million

rad

Radiation absorbed dose. The basic unit of absorbed dose of ionizing radiation. One rad is equal to the absorption of 100 ergs of radiation energy per gram of matter.

radiation (ionizing)

Particles and electromagnetic energy emitted by nuclear transformations which are capable of producing ions when interacting with matter

radioactive (decay)

Property of undergoing spontaneous nuclear transformation in which nuclear particles or electromagnetic energy are emitted

radioactivity

The spontaneous decay or disintegration of unstable atomic nuclei, accompanied by the emission of radiation

radioisotope

A radioactive isotope. An unstable isotope of an element that decays or disintegrates spontaneously, emitting radiation. More than 1,300 natural and artificial radioisotopes have been identified.

radionuclide

An unstable nuclide of an element that decays or disintegrates spontaneously, emitting radiation

reactor

A device by means of which a fission chain reaction can be initiated, maintained, and controlled; a nuclear reactor

recycle

The returning of uranium and plutonium (recovered in spent fuel reprocessing) for reuse in new reactor fuel elements

rem

Roentgen equivalent man. A dose unit which takes into account the relative biological effectiveness (RBE) of the radiation. The rem is defined as the dose of a particular type of radiation required to produce the same biological effect as one roentgen of (0.25 MeV) gamma radiation. A millirem (mrem) is one-thousandth of a rem.

The linear energy transfer, or LET, is the physical measure that specifies the relative effectiveness of equal absorbed doses from different particles in producing injuries. The higher the LET of the radiation the greater the radiation damage produced for a given absorbed dose. However, the values of LET expressed in MeV/cm are not used directly; instead the damage-producing effectiveness of a particular radiation is brought into a common scale for all ionizing radiations by the use of a quality factor (QF).

Dose equivalent (rem) = absorbed dose (rad) × quality factor (QF). The quality factor is arbitrarily assigned the value of 1 for X rays, beta particles, and gamma rays: but may be as high as 10 for alpha particles, and 30 for heavy recoil nuclei.

roentgen (R)
A measure of the ability of gamma or X rays to produce ionization in air. One roentgen corresponds to the absorption of about 86 ergs (100 ergs = 6.24×10^7 MeV) of energy from X or gamma radiation, per gram of air. The corresponding absorption of energy in tissue may be from one-half to two times as great, depending on the energy of the radiation and the chemical composition of the tissue. The roentgen is thus more useful as a measure of the amount of gamma or X rays to which one is exposed than as a measure of the dose of such radiation actually received.

reprocessing
Chemical processing of irradiated nuclear reactor fuels to remove desired constituents

short-lived nuclides
Radioactive isotopes with half-lives not greater than about thirty years, e.g., cesium-137 and strontium-90

tritium
A radioactive isotope of hydrogen with 2 neutrons and 1 proton in the nucleus

uranium
A naturally radioactive element with the atomic number 92 and an atomic weight of approximately 238. The two principal naturally occurring isotopes are the fissionable uranium-235 (0.7 percent of natural uranium) and the fertile uranium-238 (99.3 percent of natural uranium).

USAEC
United States Atomic Energy Commission (see AEC)

waste, radioactive
Equipment and materials (from nuclear operations) that are radioactive or have radioactive contamination and for which there is no recognized use or for which recovery is impractical

whole-body dose (total body dose)
The radiation dose to the entire body

Selected References

Adams, J. A. S., and Lowden, W. M., eds. *The Natural Radiation Environment.* Chicago: University of Chicago Press, 1964.

American Medical Association Council of Scientific Affairs. "Health Evaluation of Energy-Generating Sources." *Journal of the American Medical Association* 240, no. 20 (November 10, 1978).

Berger, John J. *Nuclear Power: The Unviable Option.* New York: Dell Publishing Co., 1977.

Deffeyes, K., and MacGregor, I. D. "World Uranium Resources." *Scientific American,* 242, no. 1 (1980).

Eisenbud, Merril. *Environmental Radioactivity.* New York: Academic Press, 1973.

Electric Power Research Institute, Nuclear Safety Analysis Center. *Analysis of the Three Mile Island Unit 2 Accident.* EPRI Report NSAC-1. Palo Alto, Calif.: Electric Power Research Institute, 1979.

Environmental Protection Agency. *Radiological Impact Caused by Emissions of Radionuclides into the Air in the United States.* Washington, D.C.: Office of Radiation Programs, 1979.

Florida State Department of Natural Resources, Bureau of Geology, Phosphate Land Reclamation Study. "Commission Report on Phosphate Mining and Reclamation." Tallahassee: Department of Natural Resources, 1978.

Ford Foundation and Mitre Corporation. Nuclear Energy Policy Study Group, *Nuclear Power Issues and Choices.* Ballinger Publishing Co., 1977.

Friedlander, Gerhart; Kennedy, Joseph W.; and Miller, Julian M. *Nuclear and Radiochemistry.* 1949. Reprint. New York: John Wiley and Sons, 1964.

International Atomic Energy Agency. *Radon in Uranium Min-*

ing, Proceedings of a Panel, Washington, Sept. 4–7, 1973. Vienna: International Atomic Energy Agency, 1975.

Kendall, H. W., et al. "The Risks of Nuclear Power Reactors, a Review of 'The Reactor Safety Study WASH 1400.' " Cambridge, Mass.: Union of Concerned Scientists, 1977.

Lapp, Ralph E.; Andrews, Howard L. *Nuclear Radiation Physics.* 1948. Reprint. Englewood Cliffs, N.J.: Prentice-Hall, 1963.

Leverenz, F. L., Jr., and Erdmann, R. C. "Comparison of the EPRI and Lewis Committee Review of the Reactor Safety Study." EPRI Report NP-1130. Prepared by Science Applications, Inc., for the Electric Power Research Institute Abstract. Palo Alto, Calif.: Electric Power Research Institute, 1979.

McBride, J. P.; Moore, R. E.; Witherspoon, J. P.; and Blanco, R. E. "Radiological Impact of Airborne Effluents of Coal and Nuclear Power Plants." *Science* 202, no. 4372 (December 8, 1978):1045–50.

Mongeau, S., and Poirier, H. "X-Rays and Cancer." *World Press Review,* December, 1979.

National Academy of Sciences. "Nuclear Radiation: How Dangerous Is It?" Transcript of an Academy forum, September 27, 1979, Washington, D.C.

National Academy of Sciences, Committee on Science and Public Policy. *Risks Associated with Nuclear Power: A Critical Review of the Literature.* Washington, D.C.: National Academy of Sciences, 1979.

National Academy of Sciences–National Research Council. "The Effects on Populations of Exposures to Low Levels of Ionizing Radiation, 1980." Report of the Advisory Committee on the Biological Effects of Ionizing Radiations, Division of Medical Sciences. Washington, D.C.: National Academy of Sciences–National Research Council, 1980. This report and one published in 1972 are the so-called BEIR Reports.

National Council on Radiation Protection and Measurements. *Basic Radiation Protection Criteria.* Washington, D.C.: NCRP Publications, 1971.

Nolan, W. A. "How Dangerous are X Rays?" *McCalls,* March, 1980.

O'Riordan, M. C.; Duggan, M. J.; Rose, W. B.; and Bradford, G. F. *The Radiological Implications of Using By-Product Gypsum as a Building Material.* National Radiological Protection Board, NRPB-R7, Harwell, Eng.: National Radiological Protection Board, 1972.

Pohl, R. O. "Health Effects of Radon-222 from Uranium Mining." *Search* 7, no. 8 (1976):345–50.

Rogers, V. C., and Baird, R. "Radiation Pathways and Potential Health Impact from Inactive Uranium Mill Tailings." *Transactions of the American Nuclear Society* 30 (1978):92–93.

Shapiro, Jacob. *Radiation Protection—A Guide for Scientists and Physicians*. Cambridge, Mass.: Harvard University Press, 1972.

United Nations Scientific Committee on the Effect of Atomic Radiation. "Report of the United Nations Scientific Committee on the Effect of Atomic Radiation: Sources and Effects of Ionizing Radiation." United Nations General Assembly, 32d sess. New York: United Nations, 1977.

United States, Nuclear Regulatory Commission. "*A Radiological Assessment of Radon-222 Released from Uranium Mills and Other Natural and Technologically Enhanced Sources.*" Report NUREG/CR-0573. Washington, D.C., U.S. Nuclear Regulatory Commission, 1979. Available from National Technical Information Service, Springfield, Vir.

———. "Reactor Safety Study—An Assessment of Accident Risks in U.S. Commercial Nuclear Power Plants." Report WASH 1400, also Report NUREG-75/014. Washington, D.C.: U.S. Nuclear Regulatory Commission, 1975. Available from the National Technical Information Service, Springfield, Vir.

White, David C. "Energy Choices for the 1980's." *Technology Review,* August-September, 1980, pp. 30–40.